RSPB
BRITISH
BIRDFINDER

a million voices for nature

The RSPB speaks out for birds and wildlife, tackling the problems that threaten our environment.
Nature is amazing – help us keep it that way.

If you would like to know more about The RSPB, visit the website at www.rspb.org.uk
or write to: The RSPB, The Lodge, Sandy, Bedfordshire, SG19 2DL; 01767 680551.

Published 2012 by Bloomsbury Publishing Plc, 50 Bedford Square, London WC1B 3DP

Copyright © 2012 text by Marianne Taylor
Copyright © 2012 in the photographs remains with the individual photographers – see credits on pages 287–288

The right of Marianne Taylor to be identified as the author of this work has been asserted by her in accordance with the Copyright, Designs and Patents Act 1988.

ISBN (print) 978-1-4081-5867-8

A CIP catalogue record for this book is available from the British Library

This book is produced using paper that is made from wood grown in managed sustainable forests. It is natural, renewable and recyclable. The logging and manufacturing processes conform to the environmental regulations of the country of origin.

Commissioning Editor: Nigel Redman
Project Editor: Julie Bailey
Design: Julie Dando, Fluke Art

Printed in China by C&C Offset Printing Co Ltd.

10 9 8 7 6 5 4 3 2 1

Visit www.bloomsbury.com to find out more about our authors and their books.
You will find extracts, author interviews and our blog, and you can sign up for newsletters
to be the first to hear about our latest releases and special offers.

RSPB
BRITISH
BIRDFINDER

Marianne Taylor

B L O O M S B U R Y

LONDON · BERLIN · NEW YORK · SYDNEY

Contents

Introduction

Birdwatching as a hobby is more popular than ever. Whether in the back garden, the town park or the countryside at large, more and more people are spending time watching, studying and enjoying Britain's bird life. It is a tricky business to balance human demands on the countryside with the preservation and development of biodiversity. However, a host of imaginative solutions to the problem have helped bring birds and other wildlife right into the hearts of our cities, and brought people face to face with our rarest and most iconic species.

Finding and watching

Sometimes you'll see a bird and not know what it is – that's when you need a field guide, with illustrations and descriptions that will help you to identify what you see. This book sets out to address the opposite problem, which newcomers to birdwatching will come up against sooner or later – when you know which birds you really want to see, but don't know how to find them.

To find birds in the wild you'll need a range of skills and knowledge. Some species have very exacting habitat needs, which may change quite drastically from season to season. Some are intensely shy of people, so you'll need clever fieldcraft to get a view without being seen. Others lead lives that are so strongly influenced by outside forces like weather or tidal patterns that you stand little chance of seeing them at all unless you wait for exactly the right conditions.

After the finding will come the watching, with any luck. Most of us want more than just a tick against a list, and hope for longer, closer views and the chance to observe some interesting behaviour. This book includes details on how to get the most from your sightings – ways that you can increase your chances of better views without causing disturbance, and specific interesting behaviours you might see.

Bittern

Common, scarce and rare

At the time of writing, an amazing 596 different bird species have been seen in the wild in Britain (that doesn't include species that have only occurred as escapes from captivity). A large proportion of those species are considered extreme rarities. They have only been seen once or a handful of times – they are 'accidentals', lost migrant birds from hundreds of miles away. Their future occurrence is completely unpredictable and indeed may never happen again.

At the other end of the extreme are our regular breeding and wintering birds, present in significant numbers every year. Then there are regular passage migrants, which neither breed nor overwinter here but regularly stop off in significant numbers while migrating. All of these species are covered in this book. In between the regulars and the accidentals are those species which occur in very small numbers (whether as breeders, winterers or migrants) – their presence is often unpredictable, but a few tens or hundreds turn up almost every year.

Only the more regular and more predictable of these scarce species are included in this book. Deciding where to draw the line for inclusion was impossible to manage in any truly objective way, but the species selection largely follows that used in the *RSPB Handbook of British Birds*. Of course, any birdwatcher could find a real rarity, but for species with very few British records, there is little useful advice that can be given for would-be finders, as we simply don't have enough information. As a general rule, rare birds are most likely to be found at coastal headlands in autumn, when there have been windy conditions that would push migrating birds our way.

How to use this book

This book covers nearly 300 British birds, with a full page for most species. The accounts begin with an at-a-glance summary of the species' "vital statistics", indicating how common it is on a 1–5 scale (1 being most common), its preferred habitats (see below) and when in the year it occurs.

| town and garden | woodland | open lowlands | open uplands | river and lake | marsh and estuary | at sea |

The maps use four colours, indicating when a species is present in the coloured area. The maps should be used as a guide only, as bird occurrence can be unpredictable, especially during migration periods.

 Present all year round

 Present in the breeding season only – spring to autumn

 Present outside the breeding season only – autumn to spring

 Present during migration periods only – through spring and/or autumn

The same colours with lighter tints are used to map distribution offshore for seabirds.
For a few species, hatched colouring is used to indicate that their occurrence is particularly unpredictable.

Locations of particularly good sites for some of the more uncommon birds are indicated on the maps with numbers (running top left to bottom right), and the names of those sites are listed as **Super sites** below the maps. Those that are RSPB reserves are prefixed with RSPB, while sites belonging to the Wildfowl and Wetlands Trust are suffixed with WWT, Natural England's National Nature Reserves with NNR and National Trust properties with NT.

We take a more detailed look at how to find the species through the following subsections. **Timing** looks at whether you can expect to see the bird at different times of day and year. **Habitat** gives a detailed description of the habitat/s used by the species, including seasonal variations. **Search tips** looks at habits and features that will help you to find the bird, and how its habits may change in different environments.

Watching tips looks at how to achieve longer and closer views, what kinds of interesting behaviour to look for and where, and offers advice on attracting the species to your garden. The accounts also include one or more photographs of the species as you might see them in a typical British context.

Interspersed among the species accounts are ten double-page spreads of more general birdwatching advice, divided up by theme and covering topics such as seawatching, reporting sightings of rarities, and getting the best out of garden birdwatching.

The Birdwatcher's Code

The thrill – and frustration – of birdwatching is that birds are unpredictable. They have their own lives to lead, they are subject to the vagaries of nature, and even with the most meticulous planning you'll still need a small or large dose of luck to see what you want to see. Whatever happens, though, you must never allow your desire to see a bird to trump your duty of care to the bird's welfare. This is the cornerstone of the Birdwatcher's Code, a set of recommended guidelines for all birdwatchers to follow at all times. You can view the full code at the RSPB's website.

As well as taking every care to avoid any kind of disturbance to the birds themselves, you must respect the local countryside and its byelaws, and avoid damaging habitats and property. Always set a good example to other countryside users. Another component of the code is the importance of submitting records – the BTO's BirdTrack survey provides a very quick and easy way to log your sightings, and you will help contribute to our knowledge and understanding of bird populations and migration patterns.

Bewick's Swan

Cygnus columbianus

Our smallest swan is a winter visitor to the UK and Ireland from Siberia. Most birds concentrate around the Wash, but also alongside the Severn and along the south coast. Migrants may visit suitable habitat in Scotland. Only a few hundred now winter in Ireland.

Length 115–127cm

1	2	3	4	5

J F M A M J J A S O N D

How to find

■ **Timing** Bewick's Swans arrive in the UK from mid-October, and begin to depart in March. They roost overnight on open water, and may remain there feeding during the day – they may also graze in fields at night when visibility is good. Severe winter weather may force flocks to roam away from traditional sites in search of unfrozen water.

■ **Habitat** These swans spend less time feeding on arable fields than do Whoopers, and are more likely to be found on low-lying, shallow, well-vegetated lakes, saltmarsh and damp or flooded fields. At certain wetland nature reserves large numbers are attracted to daily feeding sessions.

■ **Search tips** In the right sites, Bewick's Swans are usually easy to find, as they are highly gregarious and quite visible when feeding. Flocks of swans feeding in fields should be checked for the odd Bewick's, which have much shorter necks than Whooper Swans and, in flight, have faster wingbeats. During prolonged freezing spells, any sizeable shallow lake that is still unfrozen is worth checking. Flocks may call in flight, with rather soft, low-pitched whooping notes that are less attention-grabbing than the calls of Whooper Swans.

Super sites

1. RSPB Ouse Washes
2. Welney WWT
3. RSPB Minsmere
4. Slimbridge WWT
5. RSPB Elmley Marshes
6. RSPB Pulborough Brooks

WATCHING TIPS

From the hides at nature reserves around the Wash, you can watch large numbers of Bewick's Swans feeding and interacting at close range, without risking disturbance. Otherwise, avoid approaching flocks, as they are easily disturbed and need to conserve their energy. Within the flocks, it is possible to discern pairs and family groups. Courtship displays occur in winter. At migration time, birds travel in smaller groups then gather at wintering sites, and may stop off at almost any low-lying wetland, especially along the coasts. This is a much studied species, and you may see individuals with colour rings or neck collars.

Cygnus cygnus

Whooper Swan

This large, slim-necked, highly gregarious wild swan is a winter visitor to the northern UK, Ireland and East Anglia from Iceland and Scandinavia, though a handful of pairs breed most years in Scotland.

Length 145–160cm

| 1 | 2 | 3 | 4 | 5 |

J F M A M J J A S O N D

How to find

■ **Timing** Birds start to arrive in October, with numbers steadily building up over the months to peak in December. The return migration begins in March. Birds roost overnight on open water, moving to their feeding grounds at dawn where they will spend much of the day.

■ **Habitat** Whooper Swans roost in open water, sometimes at sea in sheltered, shallow bays. They feed on flooded pasture, shallow, well-vegetated lakes, and arable fields including cereal stubbles, potato fields and oilseed rape. At some nature reserves, regular feeding sessions attract large numbers of wild Whoopers.

■ **Search tips** When driving through suitable habitat, watch for the gleaming white of feeding Whooper Swans in the fields. Check winter gatherings of Mute Swans carefully as odd Whoopers may join them, standing out with their slighter, slimmer-necked outline and different bill and face pattern. Flocks call with noisy trumpeting notes. In summer, the occasional single or pair may be encountered at reedy lochans in highland moors; take great care to avoid disturbance. A handful fail to make their return migration, usually due to injury – such birds oversummer at the wintering site, or base themselves at a more sheltered lake. These stragglers, and feral birds from wildfowl collections, may breed in the wild.

Super sites

1. RSPB Vane Farm
2. RSPB Mersehead
3. Martin Mere WWT
4. RSPB Dee Estuary – Burton Mere Wetlands
5. RSPB Ouse Washes
6. Welney WWT

WATCHING TIPS

Watch feeding Whoopers from a good distance, as they are easily disturbed. Close views can be enjoyed from the hides at nature reserves such as Ouse Washes (Norfolk) or Vane Farm (Perth and Kinross). Both Whooper and Bewick's Swans show individual variation in their bill patterns, making it possible to identify particular individuals. Also look out for birds with wing tags or neck collars. In late winter you could see Whooper Swans performing their noisy courtship displays in open water.

Mute Swan

Cygnus olor

The commonest swan in the UK. A widespread breeding bird – non-breeders form large gatherings at traditional sites.

Length 140–160cm

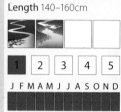

	1	2	3	4	5

J F M A M J J A S O N D

How to find it

■ **Timing** This swan can be seen at any time of day and throughout the year. Eggs are usually laid in April, with cygnets appearing in May.

■ **Habitat** Pairs favour well-vegetated lowland lakes, including those in town parks, canals or slow-moving rivers. Non-breeders flock in fields, at lakes, rivers and estuaries, with some sites attracting hundreds of birds.

■ **Search tips** A pair holding territory are usually quite obvious. As the female incubates on her large waterside nest, the male patrols nearby, often attacking other waterfowl near the nest. Birds flying overhead can be detected by their loud whooshing wingbeats.

WATCHING TIPS

Pairs in urban parks and canals are usually very tolerant of people (though care should be taken not to approach the nest) and can be studied through the year. The famous Swannery at Abbotsbury in Dorset is a remarkable example of a nesting colony of Mute Swans.

Greylag Goose

Anser anser

The largest grey goose. Lowland England has a feral population. Resident wild birds breed in north Scotland, many more visit in winter.

Length 75–90cm

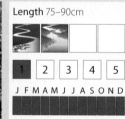

	1	2	3	4	5

J F M A M J J A S O N D

How to find it

■ **Timing** Overwintering Greylag Geese visit from September to March or April. Both residents and visitors tend to roost on lakes or estuaries and feed in fields during the day.

■ **Habitat** Many town parks hold a few feral Greylag Geese, some obviously of recent domestic origin. Larger and more rural lakes, sea lochs and reservoirs, in close proximity to suitable fields for grazing, can attract very large gatherings. Wild Scottish birds breed around remote lochs.

■ **Search tips** Feral birds are obvious across much of lowland England. Flocks on the move give themselves away with loud bugling calls. Scanning pasture and arable fields in suitable habitat should reveal feeding birds.

WATCHING TIPS

Urban feral Greylag Geese are usually very approachable, especially where they are fed (offer grain or duck pellets rather than bread). Wild birds are flighty and best observed from a distance. In winter, visiting roost sites at dawn or dusk can provide wonderful views of skeins leaving or coming into roost.

Anser fabalis

Bean Goose

Two subspecies of this grey goose visit the UK in winter. The small 'Tundra' form visits mainly the east and south-east in variable numbers. The larger 'Taiga' form is found on the Slamannan Plateau in Stirlingshire, and the Yare Valley in Norfolk.

Length 66–84cm

| 1 | 2 | **3** | 4 | 5 |

J F M A M J J A S O N D

Tundra

How to find

■ **Timing** Bean Geese arrive in the UK from mid-September, and depart in March. Cold weather on the continent pushes more Tundra Beans to our shores, but numbers of Taiga Beans are more consistent. Both subspecies spend their nights on open water, departing for the fields at dawn.

■ **Habitat** Both subspecies are more likely to be seen in fields than on water. Tundra Beans prefer to feed on arable farmland, such as stubble fields, where they will often join other grey goose flocks, especially gatherings of Pink-footed and White-fronted Geese. Taiga Beans favour lusher, marshier fields, and are more likely to associate with Greylag Geese.

■ **Search tips** With Taiga Beans, there are two core areas to search. The RSPB's Buckenham and Cantley Marshes reserve is the place to explore for the Norfolk flock. Search fields around Fannyside Lochs for the Slamannan flock. For Tundra Beans, check suitable fields for any flock of Pink-footed or White-fronted Geese, especially during cold snaps. Among Pink-footed flocks, look for birds that are slightly larger, darker and browner, with orange rather than pink legs and bill bands.

WATCHING TIPS

Probably the best way to locate either form of Bean Goose is to drive or walk around lanes in suitable habitat, scanning for feeding flocks. As with other wild geese, be very careful not to approach too closely, as the birds are wary and vulnerable to disturbance – viewing through a telescope is advisable. When seen alongside Pink-footed Geese, notice the Bean Goose's slimmer neck, longer bill and more uniform plumage. Beware occasional Pink-footed Geese with more orangey legs and Bean Geese with pink bills.

Super sites

1. Fannyside Lochs
2. RSPB Buckenham and Cantley Marshes

Taiga

Pink-footed Goose
Anser brachyrhynchus

This is the most numerous wintering grey goose in the UK, concentrating around major estuaries from the Wash northwards, straying further south in severe weather. It is extremely gregarious and vocal, and provides a wonderful spectacle when moving en masse.

Length 60–75cm

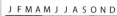

1	2	3	4	5

J F M A M J J A S O N D

How to find

■ **Timing** Pink-footed Geese arrive in the UK from their Spitsbergen breeding grounds in October, and depart in April. They roost on open water or estuarine mudflats, and at dawn leave for the fields where they will feed for much of the daytime. Dusk and dawn are the times to witness the full drama of skeins commuting to and from the roost.

■ **Habitat** Most flocks winter near the coast, where mudflats provide safe roosting grounds. Some flocks may roost on lakes or flooded fields. The birds move to fields to feed, favouring cereal fields, root crops including potatoes, and pasture. Flocks often drop into lakes to rest and bathe in the late afternoon.

■ **Search tips** In the right areas, Pink-footed Geese are easy to see, especially early and late in the day when flocks announce their approach with a loud chorus of pleasant high-pitched bugling honks. In flight the compact proportions, with small dark bill and dark head and neck, are distinctive. The large roosting flocks break up as they leave the roost and spread out over a wide area, so exploring suitable farmland and lakes within a few miles of the roosts should produce sightings. Stray singletons may join flocks of other grey geese.

Super sites

1. RSPB Udale Bay
2. RSPB Loch of Strathbeg
3. RSPB Mersehead
4. RSPB Campfield Marsh
5. RSPB Marshside
6. RSPB Snettisham
7. RSPB Berney Marshes

WATCHING TIPS

Witnessing Pink-footed Geese leaving their roost against a dramatic sunrise is one of the most exhilarating wildlife-watching experiences. As the skeins tend to radiate inland in different directions, there is no need to position yourself too close to the roost for great views as the birds pass overhead, calling constantly. Arrival at the roost at dusk is a more drawn-out affair. Flocks coming in to land perform remarkable aerobatics ('wiffling'), where they stall, sideslip and tumble downwards. Care should be taken not to disturb birds, either at the roost or by day.

Anser albifrons

White-fronted Goose

Two subspecies of this distinctive medium-sized grey goose overwinter in the UK. Birds from Siberia (*A. a. albifrons*) visit mainly the south and east coasts and the Severn Estuary, while Greenland birds (*A. a. flavirostris*) come to Ireland and west Scotland.

Length 65–78cm

1	2	3	4	5

J F M A M J J A S O N D

How to find

■ **Timing** Siberian White-fronts start to arrive from October, though peak numbers are not reached until December, and depart in March. Greenland birds stay longer, arriving in September and departing from mid-April. As with most other wintering geese, White-fronts feed in fields by day, and roost on open water.
■ **Habitat** Preferred feeding fields are arable crops such as sugar beet, potatoes and cereals, but in Scotland they will also feed on grass and clover roots in wet peaty pastures. Roosting sites include estuaries, lakes and lochs.
■ **Search tips** In England, small parties of White-fronted Geese often join larger flocks of Greylag Geese, though tend to stick together within the flock. Checking rural, low-lying areas with suitable fields, where large numbers of Greylag Geese are known to congregate, is a good strategy for finding White-fronted Geese – especially following severe winter weather. In Scotland, the Greenland birds form large gatherings, particularly when going to roost. In flight the call is higher-pitched than that of Greylag Geese, and the heavy black belly-barring stands out when quickly scanning a flock of flying Greylag Geese. They also look proportionately longer-winged and more agile in flight.

Super sites

1. RSPB The Loons and Loch of Banks
2. RSPB Loch Gruinart
3. RSPB Vane Farm
4. RSPB Ynys-hir
5. Slimbridge WWT
6. RSPB Elmley Marshes
7. RSPB Dungeness

WATCHING TIPS

To watch dramatic numbers of White-fronts, head for one of the key Scottish wintering sites. Flocks leaving or going into their roosts form an impressive spectacle. When birds first arrive, the juveniles lack the distinctive white facial blaze and dark belly-barring, and watching a feeding flock provides opportunities to study this difficult plumage, which is rather similar to the Bean Goose. Flocks sometimes visit open water in the middle of the day to drink and bathe. Young birds remain with their parents through the winter, and if you watch carefully you should be able to discern the family groups by their interactions.

Canada Goose

Branta canadensis

The Canada Goose is a common introduced bird. A handful of wild North American birds also stray to north-western coasts.

Length 90–100cm

| 1 | 2 | 3 | 4 | 5 |

J F M A M J J A S O N D

How to find it

■ **Timing** Canada Geese in the UK do not migrate and can be seen throughout the year. They usually roost on lakes, and visit fields to feed in the daytime, but spend more time on water than the grey geese.
■ **Habitat** From urban park lakes through canals, reservoirs, rivers, lakes and coastal marshland, all are likely to have a resident flock of Canada Geese.
■ **Search tips** In most parts of lowland Britain this is the commonest goose, and can be found easily in most wetland habitats. Genuine wild Canada Geese are usually found within flocks of other geese in western Scotland and Ireland.

WATCHING TIPS

These geese can be watched and enjoyed going about their daily life at close range in town parks with lakes with minimal effort. Goslings usually appear in May.

Egyptian Goose

Alopochen aegyptiaca

This small, duck-like goose, native to Africa, has been introduced to the UK and is spreading across south and east England.

Length 63–73cm

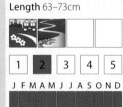

| 1 | 2 | 3 | 4 | 5 |

J F M A M J J A S O N D

How to find it

■ **Timing** The Egyptian Goose is resident, and spends much of its day on or next to water. It may roost in trees.
■ **Habitat** It generally prefers more undisturbed, well-vegetated or even wooded areas of fresh water, including large lakes, rivers and canals, though in towns may visit quite small ponds. Sometimes grazes in fields.
■ **Search tips** Usually seen in pairs or small parties, the Egyptian Goose is often quite quiet and discreet, though does give harsh duck-like calls in flight. Check quiet corners of lakes, and also look out for birds perched in waterside trees. In flight the large white wing patches are striking.

WATCHING TIPS

In urban settings such as London's Hyde Park, Egyptian Geese are confiding and approachable. You can watch courtship and territorial behaviour from late winter, while goslings appear from April. Some individuals have all-white heads.

Branta leucopsis

Barnacle Goose

Two separate populations of this pretty goose overwinter in the UK – Greenland birds on the Hebrides, especially Islay, and Spitsbergen birds around the Solway Firth. Additionally, there are many feral birds elsewhere in the UK, usually associating with flocks of other geese.

Length 58–70cm

| 1 | 2 | **3** | 4 | 5 |

J F M A M J J A S O N D

How to find

■ **Timing** Wild Barnacle Geese start to arrive in October, and leave from late March, with stragglers remaining into May. They tend to feed close to their roosting areas.
■ **Habitat** This is a goose of flat coastlines, feeding on vegetation-rich mudflats, wet pastures and arable fields. It roosts on the sea in shallow, sheltered bays or on sandbanks or islands. Feral birds may join Greylag or Canada Goose flocks well inland, on low-lying fields and lakes.
■ **Search tips** As Barnacle Geese are highly gregarious and very noisy, it is usually easy to locate feeding flocks within the right habitat at the right time of year, especially as many key feeding areas lie within designated nature reserves. In flight, groups form lines or Vs. The odd wild Barnacle Goose may be found among Pink-footed Goose flocks, away from the usual Barnacle Goose wintering grounds. You may find feral Barnacle Geese by carefully searching through large rural Greylag or Canada Goose flocks in fields or on water, looking for a smaller bird with quicker feeding motions. In some areas you may find small flocks of feral Barnacle Geese, which may or may not feed with other geese.

Super sites

1. RSPB Loch of Strathbeg
2. RSPB Loch Gruinart
3. Caerlaverock WWT
4. RSPB Mersehead
5. RSPB Campfield Marsh
6. RSPB Ynys-hir

WATCHING TIPS

It is essential that the Barnacle Goose's feeding grounds are protected from disturbance, especially as the UK flocks represent a significant proportion of the world population. You can safely enjoy good views from hides at a number of nature reserves along the Solway Firth and on Islay. Numbers are highest in the mid-winter, with gatherings reaching 10,000 at RSPB Mersehead and more than double that at RSPB Loch Gruinart. Feral Barnacles do not migrate and can be watched year round – they may breed, and may also hybridise with other feral goose species, even much larger species such as Canada or Greylag Geese. Hybrids with Barnacle Goose ancestry usually have black breasts and mostly white faces.

Brent Goose

Branta bernicla

This is our smallest and darkest goose. Two subspecies overwinter in Britain – the 'dark-bellied' *B. b. bernicla* winters along the south-east coast from Hampshire up to Lincolnshire, while the 'pale-bellied' *B. b. hrota* visits Ireland and Northumberland.

Length 56–61cm

1	2	3	4	5

J F M A M J J A S O N D

How to find

■ **Timing** Brent Geese arrive in the UK from September, and most have departed by mid-April. It usually roosts on water, often on the sea in sheltered areas or in calm weather, moving to feeding grounds by day.

■ **Habitat** The preferred feeding grounds are extensive saltmarshes with eelgrass (the main food source) and mudflats. It also feeds on open water, both salt and fresh, and flocks often fly along the coastline or just offshore between feeding areas. Brent Geese may also gather on coastal pastures.

■ **Search tips** Along suitable coastlines, look out for flocks of small, very dark geese flying past close inshore, often in lines or (less often) untidy Vs. The geese are easily mistaken for ducks at first glance, but the all-dark plumage, very small bill and proportionately long-necked outline distinguishes them from all duck species. Scan wet fields (especially dips and ditches) and saltmarsh for feeding parties, but avoid disturbing feeding or roosting birds. When flocks gather on water they are noisy, but in flight are often quiet. Autumn boat trips around flat coastlines, for example off the north Norfolk coast, can provide excellent close views of Brent Geese. Stormy weather at sea may drive birds upriver and further inland.

Super sites

1. RSPB Lough Foyle
2. Lindisfarne NNR
3. RSPB Titchwell
4. RSPB Stour Estuary
5. RSPB South Essex Marshes
6. RSPB Elmley Marshes
7. Farlington Marshes WT

WATCHING TIPS

In areas where they are somewhat used to human activity – for example around busy harbours – Brent Geese can be quite approachable. Autumn seawatches can produce high counts and close views of Brent Geese on the move. It is well worth carefully checking each bird in a flock, as one or two pale-bellied Brent Geese may be found among dark-bellied flocks and vice versa, while the odd 'Black Brant' of the North American subspecies *B. b. nigricans* turns up among flocks of both other forms. Brent Goose flocks occasionally also attract rarities such as Red-breasted Goose.

A large, colourful and distinctive duck, the Shelduck is mainly associated with flat coastlines around the whole of Britain, but may also be found inland. Numbers receive a boost in winter as continental birds arrive.

Length 58–67cm
Wingspan 110cm

| 1 | 2 | 3 | 4 | 5 |

J F M A M J J A S O N D

How to find

■ **Timing** Shelducks are present year round in the UK, but numbers fall in mid-summer as many migrate to Waddenzee in Germany to moult – there is another major moult gathering at Bridgwater Bay in Somerset, and smaller flocks in the Firth of Forth and the Wash. Elsewhere, numbers build up again through autumn.

■ **Habitat** Most Shelducks breed and winter around the coast, favouring undisturbed shallow coastlines and estuaries with sandbanks or mudflats where there is rich shallow water for feeding. Holes in loose soil or other hollows are needed for breeding. Some birds breed further inland, by wide rivers, reservoirs or gravel pits that offer suitable shorelines – in winter, they wander more widely and most open waters will have visiting Shelducks from time to time.

■ **Search tips** With their mostly white plumage, Shelducks catch the eye from a distance. Scanning mudflats, muddy shorelines or offshore sandbanks, you may see them standing on exposed mud or wading in shallow water. They can also be found swimming on open water. Often found singly or in pairs. Birds in a flock are often well spaced out, though pairs tend to breed in quite close proximity. Parents lead ducklings from the nest to sheltered water for safety.

WATCHING TIPS

In spring, Shelducks are engaged in courtship behaviour including mutual displays between the pair, which often take place on water. You may also observe antagonistic interactions between 'neighbours', as suitable sites often hold several pairs in a loose colony. A few weeks later the parents will escort their ducklings to the water – unlike most ducks, the male participates fully in care of the young. Before the young birds are fully grown, most adult Shelducks will begin their moult migration, leaving the ducklings behind. Multiple neighbouring broods band together in crèches, under the care of one or two 'aunties'.

WATCHING WINTER WATERFOWL

A flock of wild ducks, geese or swans flying into roost at dusk is a wonderful sight, and one that you can enjoy at many places in Britain and Ireland. Some species, like Bewick's Swans and Barnacle and Pink-footed Geese, come here in internationally important numbers. The same goes for the waders that form such spectacular flocks at our estuaries. Many of their main wintering grounds are fully protected nature reserves.

A numbers game

Many birds flock in winter, but none are as obvious as the gatherings of wildfowl and waders at coasts and wetlands. They are drawn together by concentrated resources in the form of good feeding grounds and safe roosting spots, and many eyes mean an approaching predator will be spotted sooner. In some places you can see flocks more than 10,000 strong – and any rarer species around will naturally be drawn to such gatherings.

The daily routine

To get the best views of these birds, you need to be aware of their behaviour and movements through the day. Most wildfowl spend the night on open water, which protects them from ground predators like foxes. For geese and swans, which spend much time feeding in fields, this means that they will arrive at the roost site as dusk falls, and leave again at dawn to return to their feeding grounds. The dawn departure tends to happen quite quickly, which makes it rather more impressive than the more protracted dusk arrival.

Coastal waders like Knots time their activities according to the tides rather than the sun. They feed on estuarine mudflats for as long as they can, but when the tide rises and covers the mud, that is their cue to find a place to roost. This may be higher up on the beach, or sometimes in fields or islands on lagoons just inland. Particularly high 'spring' tides occur about twice a month – check the dates in a tide table. Remember that roosting birds will be conserving precious energy, and watch them from a distance. If the birds closest to you show signs of restlessness and anxiety, move back and give them some more space.

Spot the alien

Many of the rare birds that turn up in Britain are found hiding within a large flock of a commoner related species. The best way to find an American Wigeon in Britain is to search through large flocks of Wigeons – the same goes for the American Green-winged Teal, which looks very similar to our own Teal. It is always worth checking through flocks for the odd one out.

Dunlin

Dunlin

Whooper Swans

Where to watch

- RSPB Snettisham
  (Knots, Dunlins,
  Pink-footed Geese)
- RSPB Mersehead
  (Barnacle Geese)
- RSPB Ouse Washes
  (Whooper Swans,
  Bewick's Swans)
- RSPB Berney Marshes
  (Golden Plovers)
- RSPB Morecambe Bay
  (Knots, Dunlins)
- RSPB Loch of Strathbeg
  (Pink-footed Geese)
- RSPB Loch Gruinart
  (Barnacle Geese)
- RSPB Vane Farm
  (Whooper Swans,
  Pink-footed Geese)
- RSPB Lough Foyle
  (Brent Goose)
- Slimbridge WWT
  (Bewick's Swans)
- Caerlaverock WWT
  (Barnacle Geese)
- Welney WWT
  (Whooper Swans)

Surveying the scene

Because Britain is so important for its wildfowl and wader populations, all birdwatchers are encouraged to take part in the BTO's Wetland Birds Survey (WeBS). Volunteers who sign up for the survey make regular visits through the winter to designated sites and count all wildfowl and waders seen. The data gathered by this survey helps identify which sites are the most important for which species, and whether there are any changes in numbers or distribution from year to year. You can find out more about WeBS by visiting the BTO's website.

Counting large flocks of birds can be quite difficult, especially when the flock is in flight. One technique often used is to count just a small part of the flock, say 20 or 50 birds, and then estimate how much larger the flock is than the part that you have counted. Today, many researchers will have a camera of some sort – if you can take a photo of the whole flock, you can make a reasonably accurate count of the birds later at your leisure.

Smew

Mandarin Duck

Aix galericulata

Native to China, this duck is now established as a feral species in Britain, originating from escapes from wildfowl collections. Its stronghold is the Surrey and Berkshire area but has spread to neighbouring counties, while other separate populations exist elsewhere.

Length 41–49cm

1	2	3	4	5

J F M A M J J A S O N D

How to find

■ **Timing** Mandarins can be seen at any time of year, but males are at their magnificent best in winter and early spring – this is also the easiest time to see them. Courtship displays can be seen at this time, while ducklings are usually around from late May.

■ **Habitat** Its habitat in the UK is primarily sheltered waters within woodland, such as tree-lined slow-flowing rivers and canals, and quiet park lakes. Because it nests in tree holes, mature trees need to be present close to suitable water to encourage breeding (it will also use nest boxes). Although large parks such as Windsor Great Park have free-flying Mandarin Ducks, be aware that some parkland Mandarin Ducks may be part of a captive wildfowl collection.

■ **Search tips** Despite the male's spectacular plumage, Mandarin Ducks can be shy and discreet, especially in the breeding season. Search the margins of suitable lakes and rivers for birds on the water or the shore, and also scan horizontal tree branches close to the water for perched birds, especially on islands. Larger tree holes may hold nesting females, with the male usually in attendance nearby. In areas with a lot of human traffic, Mandarin Ducks can become confiding and will come to the bank to be fed.

Super sites

1. Cannop Ponds
2. Stockgrove Country Park
3. Windsor Great Park
4. Connaught Water
5. Bough Beech Reservoir

WATCHING TIPS

The fascinating courtship display can be seen in late winter, with males posturing to the females on the water and the shore, showing off their elaborate plumage to its best advantage. Pick a site that's popular with the general public to enjoy closer views of these beautiful ducks going about their daily business. If unsure whether you are looking at captive or feral Mandarin Ducks, check the wing-tips. Captive birds have a wing-tip missing, where the primary feathers have been cut to prevent straying.

Anas penelope

Wigeon

A medium-sized dabbling duck, the Wigeon visits most of the UK in winter, especially by the coast, and breeds in much smaller numbers, primarily in Scotland. It forms very large flocks in suitable habitat, often greatly outnumbering other duck species.

Length 45–51cm

1	2	3	4	5

J F M A M J J A S O N D

How to find

■ **Timing** Birds start to arrive in earnest from October, and most will have departed by mid-April. Breeding birds tend to winter near the breeding grounds. Through the day flocks may move around from fields to shorelines or open water and back, sometimes dictated by the tides.

■ **Habitat** The largest numbers gather on low-lying coastlines, on flat landscapes with good all-round visibility, where they feed on arable fields, wet pasture and mudflats. They will also feed, rest and bathe on open water. Inland lakes and reservoirs with fields around will also attract flocks. Breeding birds base themselves around shallow, quite well-vegetated and undisturbed lochs, with suitable wooded or densely heathy ground nearby for nesting.

■ **Search tips** Wigeons are more often found on land than water, sometimes in the company of geese, which have similar feeding preferences. Scan for dense gatherings of mixed grey (male) and brown (female) ducks, working their way steadily across a field. The male's distinctive and emphatic whistled 'wheeoo' call carries long distances and often gives away the presence of a flock on the ground or flying overhead. Flocks in flight are often densely packed, with the white bellies of the females and the white shoulder-patches of the males very striking. Check sheltered bays for birds resting on the sea.

Super sites

1. RSPB Udale Bay
2. RSPB Belfast Harbour
3. RSPB Ynys-hir
4. Slimbridge WWT
5. RSPB Otmoor
6. RSPB Ouse Washes
7. Cley Marshes WT
8. RSPB Buckenham and Cantley Marshes
9. RSPB Pulborough Brooks

WATCHING TIPS

Unless you can watch from a hide, be careful not to approach feeding flocks too closely – they are alert and easily disturbed. Many hides on coastal reserves offer the chance of good views. Some sites offer the chance to watch birds at work and at rest, where wet grassland for feeding adjoins open water. In late winter you may observe courtship behaviour, including displays and ferocious chases between rival males. Avoid searching for nest-sites around the lochs, but look out on the lochs themselves for females with ducklings in early summer.

Gadwall

Anas strepera

This dabbling duck is a rather scarce breeding bird in the UK, but is much more numerous in winter. Its stronghold is the south and east of England, where its population may be derived from feral birds.

Length 46–56cm

1	2	3	4	5

J F M A M J J A S O N D

How to find

■ **Timing** The Gadwall is easier to find in winter than summer. Numbers build up through autumn, and start to fall through March.

■ **Habitat** It prefers shallow, quiet and well-vegetated low-lying lakes, as well as slow-flowing rivers. It may visit park lakes in quite urban settings. In winter it uses a wider range of water bodies, including marshes and more exposed reservoirs.

■ **Search tips** Having a rather dull plumage, the Gadwall is easily overlooked among Mallards and other wildfowl, so scan suitable lakes carefully, including the open water, the shore and any islands, checking for the males' distinctive solid black rear ends and, when on land, the females' white bellies. It tends not to form very large concentrations, and you may find a pair or two lurking among other ducks – it is also often seen among gatherings of Coots. The female rears her ducklings alone, and is very Mallard-like in appearance (as are the ducklings) – check any apparent female Mallards with ducklings carefully as numbers of breeding Gadwalls are on the increase.

WATCHING TIPS

Gadwalls tend to go about in small groups or (especially from late winter) pairs, with courtship behaviour in evidence through winter and spring. This is the best time to look for them and also to enjoy the subtle beauty of the male's breeding plumage at its best – birds established on park lakes may be quite approachable. Sometimes two or more males and females join together for communal displaying sessions. When foraging on open water they are drawn to feeding Coots, and often steal the waterweed that the Coots bring up from under water. Ducklings appear in May.

Teal

Our smallest dabbling duck, the Teal is also one of our commoner species in winter, though it is a rather uncommon breeding bird. The breeding population has its stronghold in Scotland and northern England – it is very widespread in winter.

Length 34–38cm

1	2	3	4	5

J F M A M J J A S O N D

How to find

■ **Timing** Wintering Teals start to arrive from eastern Europe and Asia in mid-autumn, their westward spread across the UK strongly influenced by weather. They begin to depart from mid-February.

■ **Habitat** Almost any lowland lake or area of marshland is potential habitat for wintering Teals, though well-vegetated shallow waters with muddy shorelines are likely to attract more birds and hold them for longer. Most breeding birds are found in uplands, nesting in scrub or heather often some distance from the tarns or lochans where they will bring their ducklings.

■ **Search tips** The small size and (from a distance) lack of strongly patterned plumage makes the Teal an inconspicuous duck – it is also usually shy and easily disturbed. Scan along the margins of lakes to look for small parties roosting on the shoreline or feeding close to the shore. The agile flight and rapid wingbeats mean flying Teals can look quite unlike a duck at first glance – look out for tight, flickering flocks circling around suitable water bodies, and listen for the whistled '*prip-prip*' call of the males. Breeding Teals are discreet – it is best not to search for nests, but by early summer females and ducklings will be on ponds or other water bodies.

WATCHING TIPS

Because Teals are rather shy and tend to keep close to the shore, the best way to watch them is from a hide. With patience you may enjoy good views of the birds feeding close at hand as they work their way around the edge of a lake. They are quick to take flight when disturbed, which may reveal the presence of a predator. Drakes will be courting from as early as November, sometimes in communal courtship displays that involve much antagonism between males, and also pursuit flights. Most birds will be paired up by March.

Pintail

Anas acuta

One of our rarest breeding ducks, the Pintail is primarily a winter visitor, especially to coastal sites, with the largest numbers around the Wash, the Solway and the Dee Estuary. It breeds sporadically in well-separated locations across the UK.

Length 51–62cm

1	2	3	4	5

J F M A M J J A S O N D

How to find

■ **Timing** Wintering Pintails from eastern Europe and Asia begin to arrive in September, and peak in December. Cold weather will push them further west. The return migration gets under way from late February.

■ **Habitat** This duck prefers large coastal wetlands, such as estuaries, saltmarsh, lagoons, wet pasture and shallow bays. It may also feed on arable land close to suitable water. It is an uncommon visitor to inland lakes and reservoirs. It favours expansive wet grasslands with shallow pools for breeding.

■ **Search tips** On the water, the high-riding, elegant outline of both male and female Pintails is eye-catching among other ducks, and they will readily swim out well away from the shore, meaning that even when only one or two birds are present at a site they are often easy to find. As well as the water, scan the shore and islands, looking out for the males' dark head and white breast and belly. In flight, Pintails stand out among other ducks with their long necks and very pointed wings. The preference of breeding birds for rather transitional habitat types means that year on year new sites may be used and old ones abandoned, making evidence of breeding difficult to find.

Super sites

1. RSPB Mill Dam
2. RSPB Mersehead
3. RSPB Saltholme
4. RSPB Dee Estuary – Parkgate
5. RSPB Ouse Washes
6. Slimbridge WWT
7. RSPB West Sedgemoor

WATCHING TIPS

Visit marshlands and lakes around major estuaries (especially the Wash, Dee and Solway) to see large numbers of these very beautiful ducks, which look at their best from November into spring. Courtship behaviour is ongoing through the winter and into spring, males performing various ritualised displays towards females and each other, though showing more tolerance of their own sex than most dabbling ducks do. At some reserves where wild swans are fed in winter, Pintails may take advantage of the food offered and you can enjoy very close views of them from hides.

Anas querquedula

Garganey

The Garganey is our only dabbling duck that is a summer visitor, most migrating south of the Sahara for winter. It is uncommon and rather unpredictable, with most seen in south-east England, and is very discreet when breeding.

Length 37–41cm

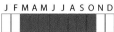

1	2	3	4	5

J F M A M J J A S O N D

How to find

■ **Timing** Garganeys are early arrivers, a few reaching the UK by late February. They depart through August and September, most having left by mid-October. They are easiest to see when they first arrive, and when young birds are dispersing in late summer and early autumn.
■ **Habitat** Prefers sheltered, shallow and well-vegetated but not shady waters, including quite small ponds and ditches, and temporary floods. It may visit wider and more coastal lakes on migration, but moves to smaller waters to breed.
■ **Search tips** Searching open water and shorelines of quiet lakes along the south and east coasts in early to mid-spring may produce sightings of new arrivals – most likely a pair or two which may stay a week or more before moving on to breeding grounds. These same lakes may also host birds in late summer and autumn, when identification presents more of a challenge as pairs have separated and males are in eclipse. At this time of year they may associate with Teals. Among Teal flocks look for a slightly larger but slender duck with a well-marked face and proportionately longer bill than Teal. When breeding, Garganey are hard to find as suitable habitat changes from year to year; best not to search for them, to avoid disturbance.

Super sites

1. RSPB Titchwell
2. RSPB Minsmere
3. RSPB Elmley Marshes
4. Stodmarsh NNR
5. Rye Harbour
6. RSPB Dungeness

WATCHING TIPS

Garganeys pair up before they arrive in the UK, so spring sightings are usually of well-bonded pairs that stick close together and may be seen engaged in courtship behaviour or mating. Autumn Garganeys – a mix of youngsters, adult females and eclipse males, present a real identification challenge, but as they are often seen alongside Teals this offers the opportunity to study how the two species differ in plumage, shape and behaviour – in fact the Garganey is more closely related to the Shoveler than the Teal.

Shoveler

Anas clypeata

This very distinctive dabbling duck breeds rather patchily across most of England and parts of the rest of the UK, but becomes much more numerous and widespread in winter when migrants from further north and east arrive.

Length 44–52cm

1	2	3	4	5

J F M A M J J A S O N D

How to find

■ **Timing** As with most other dabbling ducks, winter is the best season to look for Shovelers, as they are most numerous then and the drakes are at their best. After breeding, moulting birds get together and form small flocks.

■ **Habitat** Shovelers use a wide range of shallow lowland waters, including brackish lagoons, reservoirs, gravel pits and flooded fields, needing reasonably large expanses of open water for feeding. They will visit and even breed on town park lakes in some areas.

■ **Search tips** Shoveler males are colourful and eye-catching, though their rather low stance in the water and steady feeding style with the bill immersed and little or no upending means their movements don't particularly draw the eye. Carefully check islands and shorelines for resting birds – the males' white chests and chestnut flanks are eye-catching but sleeping females (with the characteristic bill hidden) could easily be overlooked among female Mallards. Rarely forms large flocks but lakes that are generally attractive to dabbling ducks are likely to hold a few. In flight the wings look set well back and this, coupled with the large bill, gives a strikingly front-heavy look.

WATCHING TIPS

In winter, Shovelers tend to form small parties which demonstrate interesting behaviour, including cooperative feeding where the group swims in circles, stirring up their invertebrate prey, and communal courtship, where several males get together and show off their plumage to a female with head-bobbing and tail-lifting displays. Groups may also take to the air for courtship flights, showing surprising agility in flight. Shovelers can be quite confiding, especially in areas with a lot of human visitors. In autumn, the males in eclipse exhibit a confusing array of intermediate plumages, though are easy to identify provided the bill is on view.

Anas platyrhynchos

Mallard

Our commonest and most familiar dabbling duck, the Mallard can be found on almost all kinds of water.

Length 50–65cm

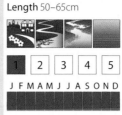

| 1 | 2 | 3 | 4 | 5 |

J F M A M J J A S O N D

WATCHING TIPS

Because Mallards are so common and confiding, it is easy to make close observations of them year-round. Ducklings may be seen right through spring and summer, while courtship behaviour is evident through winter and spring.

How to find it

■ **Timing** There is a large Mallard population in the UK year-round, with extra birds arriving from the continent in winter.

■ **Habitat** Almost any form of water attracts Mallards, from garden ponds and small streams to all lake types in both lowlands and uplands; they may also visit sheltered sea bays. It may nest a mile or more from water.

■ **Search tips** Usually very easy to find. Most town ponds have a few Mallards, often very tame and confiding. They may be joined by domestic Mallards, which may look different in both plumage and shape/size to 'wild-type' birds.

Aythya fuligula

Tufted Duck

Our commonest diving duck, the Tufted Duck breeds across most of the UK, and many more visit for the winter.

Length 40–47cm

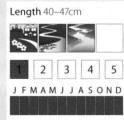

| 1 | 2 | 3 | 4 | 5 |

J F M A M J J A S O N D

How to find it

■ **Timing** Winter is the best time to look for Tufted Ducks, though the species is usually easy to find at any time of year.

■ **Habitat** Most kinds of still lowland water attract this duck, with reservoirs and gravel pits particularly good places to look.

■ **Search tips** In winter Tufted Ducks form large flocks which often sit out in the deeper parts of the lake, the black and white males particularly eye-catching over considerable distances. Check larger town park lakes, particularly in colder weather.

WATCHING TIPS

In town parks Tufted Ducks can be as confiding as Mallards and will come to the shore for food, giving opportunities to watch them swimming under water at close range. They are more shy when breeding, but some do nest in urban environments. Courtship displays are evident in late winter.

Pochard

Aythya ferina

The handsome Pochard is a rather uncommon but widespread breeding bird, and a common winter visitor – in most areas it is the second most common diving duck after Tufted Duck, and often frequents the same areas as that species.

Length 42–49cm

| 1 | 2 | 3 | 4 | 5 |

J F M A M J J A S O N D

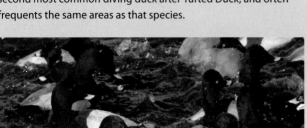

How to find

■ **Timing** As with most wildfowl, Pochards are easiest to find in winter, but are present throughout the year. Cold weather pushes them further west and towards smaller water bodies, and also to town park lakes where the water is more likely to remain unfrozen.

■ **Habitat** As with Tufted Duck, this is a bird of inland, low-lying lakes and reservoirs, and needs clear stretches of water around 1–2 metres deep for feeding. Large gravel pits may hold large flocks in winter. For breeding, it needs water bodies with well-vegetated margins, though not in wooded areas, within which it will nest, close to the shore.

■ **Search tips** In suitable locations it can form very large flocks, and is quite obvious, usually keeping well out from the margins and diving frequently when feeding. In other areas look out for odd Pochards swimming or sleeping among groups of Tufted Ducks, standing out from the latter even in poor viewing conditions by their different head shape. Breeding Pochards choose more undisturbed lakes and are very discreet until the ducklings hatch, when the females escort their broods out onto the open water.

WATCHING TIPS

The large winter flocks in the UK tend to be dominated by males, which reach the best wintering grounds first, forcing females to migrate further south on average. Good places to see very large gatherings include Abberton Reservoir in Essex, and the lakes by the Ouse Washes, but most good-sized lakes of the right depth will attract some birds, unless they become completely frozen over. In town parks Pochards will become as confiding as Mallards and Tufted Ducks, readily coming close to the shore for food and affording the opportunity for close and prolonged views.

A large diving duck, the Scaup is closely related to the Tufted Duck and Pochard but is much more coastal in its distribution. It is a winter visitor to the UK, though has nested here in tiny numbers.

Length 42–51cm

| 1 | 2 | **3** | 4 | 5 |

J F M A M J J A S O N D

How to find

■ **Timing** Scaups arrive in October and depart in March, and are best searched for in mid-winter.

■ **Habitat** The only duck in the *Aythya* genus with a strong tie to the sea, the Scaup's winter habitat includes both sheltered and not so sheltered seas as well as coastal lakes (both fresh and saltwater). At sea, its presence is largely dictated by access to mussel beds where it feeds, so its movements are influenced by the tides. Many of the larger bays and estuaries in Scotland have regular wintering flocks. Some sites that formerly attracted many Scaups because of waste grain outflow (from breweries and distilleries) have lost their flocks following 'clean-up' operations.

■ **Search tips** Visiting the best sites in mid-winter should produce sightings. Scan the water carefully, bearing in mind that birds may be temporarily out of view when diving or between wave crests. Coastal lakes are also worth searching, especially following stormy weather at sea that may force birds inland, sometimes well inland where they will seek large water bodies. Away from the sea, beware of confusion with female Tufted Duck, which may show a white facial blaze similar to female Scaup – check head shape and other features to confirm identification.

Super sites

1. RSPB Udale Bay
2. RSPB Loch of Strathbeg
3. Bridgend Bay on Islay
4. Firth of Forth
5. Loch Ryan
6. RSPB Mersehead
7. Southerness Point

WATCHING TIPS

Watching birds at sea requires patience and often good optics as well, as they may be far offshore and viewing conditions challenging. Making repeated visits will help you to gauge how the birds' activity changes with the tides. If you can find an elevated viewpoint that should help with the problem of 'losing' birds among wave crests. When Scaups turn up on coastal lakes there is the potential for more prolonged observation, and also comparison with the main confusion species – Tufted Duck. Be aware that male hybrids between Pochard and Tufted Duck are occasionally seen and can look similar to Scaup.

Red-crested Pochard

Netta rufina

The Red-crested Pochard has become established in south-central England, following escapes from wildfowl collections.

Length 53–57cm

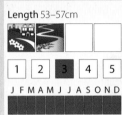

1	2	3	4	5

J F M A M J J A S O N D

How to find it

■ **Timing** This duck may be seen with equal ease at any time of year – there is no clear evidence of a winter influx.

■ **Habitat** Most UK Red-crested Pochards live and breed on larger park lakes or well-vegetated reservoirs.

■ **Search tips** Once you have identified a water body where the species occurs, it is easy to observe, often feeding well out in the middle of the lake, sometimes in company with Pochards and/or Tufted Ducks. Urban or semi-urban lakes, even those in central London parks, are good places to try.

WATCHING TIPS

Red-crested Pochards are often approachable and easy to watch. Genuinely wild Red-crested Pochards from Europe may visit the UK – any bird found away from known naturalised populations, especially on the south or east coast, may be wild rather than feral.

Ruddy Duck

Oxyura jamaicensis

Introduced from North America, this diving duck has been subject to an eradication programme and is now uncommon.

Length 35–43cm

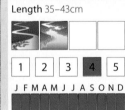

1	2	3	4	5

J F M A M J J A S O N D

How to find it

■ **Timing** Ruddy Ducks are generally sedentary and may be observed at any time of year, though may wander short distances in winter.

■ **Habitat** Prefers rather shallow lakes and pools with plenty of submerged vegetation.

■ **Search tips** Its densest populations were in the West Midlands, northern England, Anglesey and southern Scotland. Following the eradication programme, the Ruddy Duck is rare in the UK and its populations scattered. Scan suitable lakes for birds actively feeding or sleeping out on deeper water, sitting low in the water when active, but showing a long upward-angled tail when asleep. It is gregarious in winter and formerly gathered in sizeable flocks on suitable lakes.

WATCHING TIPS

The government-led eradication programme is intended to stop Ruddy Ducks spreading to Spain where they may hybridise with the endangered native White-headed Duck (*Oxyura leucocephala*). The species is likely to be closely controlled in the future if not eliminated altogether.

Somateria mollissima

Eider

One of our commoner seaducks, the large and distinctive Eider may be seen all year around the coast of Scotland, the far north of England and northern Ireland, and wanders further south in winter (though it remains most numerous in the north).

Length 50–71cm

| 1 | **2** | 3 | 4 | 5 |

J F M A M J J A S O N D

How to find

■ **Timing** Eiders are more widespread in winter and gather in flocks, though they are perhaps less predictable in where they may be found. In late winter the drakes look their best, while females with ducklings may be seen in late spring.
■ **Habitat** The Eider is rarely seen inland, though it may come quite close inshore and into bays or marine lakes depending on the tides, the weather, and where and when its underwater food supply (primarily mussels) is most accessible. It feeds mainly at depths of about 3 metres, so avoids seas where the floor slopes steeply away. It nests on sheltered and undisturbed low rocky coastlines, close to the sea.
■ **Search tips** Whenever you are at the coast, keep an eye out for Eiders swimming offshore or flying over the sea, though the further north you are the better chance you have of success. Pick an elevated viewpoint from which to scan, especially with rougher seas. Almost any coastline in Scotland should give sightings, while in winter a seawatch of a few hours elsewhere in the UK (especially the east coast of England) may well produce an Eider or two, along with many other species. Females with ducklings may be very close inshore; try small, quiet, rock-fringed bays.

Super sites

1. RSPB Mousa
2. RSPB Culbin Sands
3. RSPB Rathlin Island
4. Farne Islands NT
5. RSPB Coquet Island
6. RSPB Titchwell

WATCHING TIPS

There are several sites where in late spring and summer you can enjoy very close views of female Eiders on the nest or tending their young ducklings, such as the Farne Islands off Northumberland. Visiting bays and harbours in Scotland in late winter and early spring should allow you to watch the males' energetic and noisy efforts to court the females. Birds feed most actively on an ebbing tide, and rough weather at sea may push them closer inshore. Be aware that some quite informal wildfowl collections have captive Eiders – any Eider seen away from the coast is likely to be captive or feral.

Long-tailed Duck

Clangula hyemalis

This very beautiful small seaduck is most numerous around the Scottish coast, but may turn up further south and inland. It does not breed in the UK but can be seen in all months, though winter is best.

Length 40–47cm

1	2	3	4	5

J F M A M J J A S O N D

How to find

■ **Timing** It is best looked for in the winter months, though a few non-breeding birds can be found off Scottish coastlines in summer.

■ **Habitat** Suitable coasts are fairly shallow and sheltered, with firths, bays and harbours on the east coast of Scotland particularly good places to look. Though it is primarily a seaduck, the odd Long-tailed Duck will sometimes visit large lakes and reservoirs inland – most commonly near the coast but sometimes miles inland, and these individuals may remain at such sites for days or weeks.

■ **Search tips** The Long-tailed Duck's small size and habit of making frequent and prolonged dives can make it a challenge to find even when you are in exactly the right place. Take time to scan the sea from a high vantage point in suitable spots, especially when other seaducks are present, indicating good feeding conditions. Regular checking in winter of large water bodies, especially those near the coast, could pay off – look out for a small and compact pale duck which dives frequently and athletically, and tends not to associate with other ducks. Males in full winter plumage are very striking, but the inland wanderers tend to be of the more nondescript 'female-type'.

Super sites

1. RSPB Brodgar
2. RSPB Hobbister
3. RSPB Coll
4. Ythan Estuary
5. St Andrews Bay
6. RSPB Dungeness

WATCHING TIPS

Patience is needed to watch actively feeding Long-tailed Ducks, because they dive so often and for so long. The dive usually begins with a jump clear of the water with the tail fanned, particularly striking in the long-tailed adult males. Winter flocks show much activity, with communal courtship among the males, and aggression from paired males towards would-be usurpers. This duck has the most complex sequence of adult plumages of any British bird, with some feathers moulted three times a year, and watching flocks should reveal some of the variety on view at any given time.

Melanitta nigra

Common Scoter

This is perhaps the archetypal seaduck, although it is also an inland breeding bird in the far north-west of Scotland, albeit a very rare one. In winter it may be seen along most UK coastlines, especially in the east.

Length 44–54cm

| 1 | **2** | 3 | 4 | 5 |

J F M A M J J A S O N D

How to find

■ **Timing** Common Scoters may be seen offshore at any time of year, especially in the far north, but you are most likely to be successful in winter. The distance that flocks feed from the shore depends on the water depth so is partly determined by the tides.

■ **Habitat** It is most likely to be seen in shallow seas, off flatter shores, and rarely closer than half a kilometre from the shore. Exceptionally, odd birds may stray upriver or inshore, where they will seek large, deep reservoirs or other water bodies. It breeds in the remote Flow Country of north Scotland, nesting some distance from water among thick heathy vegetation. The ducklings are escorted to lochans when they hatch.

■ **Search tips** As with other seaducks, it is always worth scanning the sea from time to time for Common Scoters, especially in winter. Look for a tightly packed raft of buoyantly swimming, very dark, chunky ducks. Closer views will reveal the upward-angled, pointed tails, the yellow on the males' bills and the pale cheeks of the female. In flight flocks form trailing, untidy lines. After sea storms, check tidal rivers and suitable large lakes for Common Scoters and other seabirds.

Super sites

1. Moray Firth
2. RSPB Culbin Sands
3. RSPB Mersehead
4. Cardigan Bay
5. RSPB Titchwell

WATCHING TIPS

Because they tend to stay some distance from shore, Common Scoters are not easy to watch, and powerful optics will usually be necessary to discern details of plumage or behaviour. It is worth taking your time to scan flocks carefully, checking for other scoter species. Look out for the communal courtship display in late winter, whereby a party of males display, with neck and tail stretching, and call to a single female. Because this is such a rare breeding bird, and its nesting habitat is fragile and vulnerable, it is best not to search for birds on the breeding grounds away from designated nature reserves.

Velvet Scoter

Melanitta fusca

The Velvet Scoter is closely related to the Common Scoter, and is similar to it in most respects. It does not breed in the UK but is a winter visitor to inshore waters, especially down the east coast.

Length 51–58cm

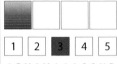

| 1 | 2 | 3 | 4 | 5 |

| J | F | M | A | M | J | J | A | S | O | N | D |

How to find

■ **Timing** Birds start to be seen from October, but the highest numbers are not reached until January. Return migration begins in mid-March.

■ **Habitat** It has similar habitat requirements to the Common Scoter, and is most likely to be found on open, exposed and relatively shallow seas, rather than in more sheltered bays – it often feeds in shallower water than Common Scoter. As with Common Scoter it may be storm-driven inland, into estuaries or further upriver, and may seek refuge on deep reservoirs or lakes. It is probably more susceptible overall than Common Scoter to being pushed inland, although the actual numbers found inland will be lower as it is a scarcer bird.

■ **Search tips** Velvet Scoters may be found among Common Scoter flocks, though patience is necessary to pick them out. The most obvious difference from any distance is the Velvet Scoter's white wing-bar – evident in flight, but if you are watching a resting flock wait for birds to stand up and wing-flap (which they do quite frequently) to check for this. Good views are necessary to discern the male's white eyepatch and female's different head pattern.

Super sites

1. Moray Firth
2. St Mary's Island
3. Filey Bay
4. RSPB Titchwell

WATCHING TIPS

Velvet Scoters may prove a little easier to watch than Commons because they feed on average a little closer inshore, and are also more confiding – small scoter flocks seen rather close inshore may prove to be Velvets rather than Commons. When Velvet and Common Scoters are together, take the opportunity to study the more subtle differences between the species. Wintering flocks will be engaged in courtship behaviour, in order to establish pairs before the return migration to the breeding grounds. Like Common Scoters, there is a communal courtship display, as well as mutual displays between pairs that are already bonded.

Goldeneye

The distinctive Goldeneye is a rare breeding bird, restricted to lochs and rivers in the central Scottish Highlands. Many more visit the UK in winter and may be found on suitable water almost anywhere, though often in small numbers.

Length 42–50cm

1	2	3	4	5

J F M A M J J A S O N D

How to find

■ **Timing** Breeding birds can be seen in the Highlands in late spring and summer. From August the birds begin to travel to their winter quarters, and remain there until March or April.

■ **Habitat** These diving ducks require fairly deep, still or slow-flowing water without too much floating or submerged vegetation. They also nest in holes, so breeding lochs and rivers must have tall trees around. In winter, they find suitable feeding grounds on lakes, reservoirs, bays, sheltered seas and rivers.

■ **Search tips** Females with ducklings are often easy to spot on their lochs from mid-summer, though the male of the pair will have departed during the incubation stage. In winter, check suitable waters for Goldeneyes, concentrating on deeper water, out from the shore, bearing in mind that feeding birds are under water for much of the time. Males look strikingly white from any distance, females a colder grey on the body than most other diving ducks. They sit low in the water, and are not especially gregarious with their own kind or other ducks when feeding, though may form large flocks when roosting.

Super sites

1. River Spey
2. RSPB Loch Garten
3. RSPB Vane Farm
4. RSPB Leighton Moss
5. RSPB Fairburn Ings
6. RSPB Valley Wetlands
7. RSPB Conwy
8. RSPB Dungeness

WATCHING TIPS

When feeding, Goldeneyes make frequent long dives, so be patient to get a good look at the bird. When both sexes are present at a site in late winter you may observe the drake's energetic courtship display, whereby the head is stretched forward then thrown over the arched back, while the feet paddle furiously to kick up a spray of water behind. Newly hatched ducklings have to exit their tree-hole or nest box with a leap into the water below, an amazing sight but all over very quickly. Females with ducklings are not especially nervous and you should have good views from the shore.

Smew

Mergus albellus

This delightful small sawbill is much sought after by birdwatchers. It is an uncommon winter visitor to the UK, mainly the south-east, and a fairly reliable visitor to some sites, though often not until very cold weather in mid-winter.

Length 38–44cm

| 1 | 2 | **3** | 4 | 5 |

J F M A M J J A S O N D

How to find

■ **Timing** Most Smews reach the UK no earlier than December, and will have departed for their breeding grounds in north-east Europe and beyond by mid-March. Severe weather on the continent pushes more birds across to the UK.

■ **Habitat** It favours fairly deep waters with little water vegetation, often close to the coast. Newly arrived or wandering Smews may drop in on quite small ponds and even streams, although large lakes, reservoirs and gravel pits are the likeliest places to find them. Lakes that hold Goldeneyes are often attractive to Smews, and vice versa.

■ **Search tips** Any water body within its usual winter range is worth checking, especially following cold weather on the continent. The most reliable sites often draw in small parties of Smews, which will stay for days or weeks providing conditions remain suitable, though if several suitable lakes are in close proximity the Smews may commute between them. Adult males are very white and quite eye-catching, but most Smews in the UK are the much less striking 'redheads' – females or first-year birds. They are rather grebe-like in general appearance and behaviour – scan open, deeper water for a small, low-swimming bird that dives often and energetically.

Super sites

1. RSPB Loch of Strathbeg
2. Wraysbury Gravel Pits
3. Abberton Reservoir
4. Rye Harbour Nature Reserve
5. RSPB Dungeness

WATCHING TIPS

Most sightings of Smews in the UK will be birds that are busy feeding, replenishing their energy stores after a long journey. You may therefore have the challenge of keeping track of a small bird that dives often and travels considerable distances under water, though note that resting birds often move onto dry land, where they may be very inconspicuous. If you are lucky enough to find a group of Smews that includes multiple adult males you may observe the communal courtship display, especially in the afternoons and later in winter. Both sexes will perform displays, involving posturing and fast swimming.

Mergus serrator **Red-breasted Merganser**

This sawbill breeds patchily in Wales, Scotland, Ireland and north-west England, having spread south in recent years. Its numbers here increase four-fold in the winter, when it visits the whole UK coastline, favouring large bays and estuaries.

Length 52–58cm

1	2	3	4	5

J F M A M J J A S O N D

How to find

■ **Timing** Breeding Red-breasted Mergansers start to arrive on their nesting grounds from March, and head back towards coastal wintering sites from July, being joined by more from the continent through the autumn.

■ **Habitat** It breeds in sheltered spots mostly close to sandy coastlines or to running water, though sometimes also by quiet inland lakes, with plenty of vegetation including trees in the vicinity. Post-breeding, may gather in flocks in river mouths. In winter, most are at sea, in relatively shallow water, but it also visits lakes and reservoirs, especially on migration.

■ **Search tips** Watch the sea from high on the beach for Red-breasted Mergansers, bearing in mind that they are as likely to be seen in flight, usually low over the waves, as swimming on the sea. Look for a rather untidy and not very duck-like outline, long-bodied and thin-billed with a straggly crest. Although this species is highly gregarious and forms quite large flocks, it is by no means unusual to see singles or small parties. On suitable breeding habitat, broods tend to aggregate together – look for large gatherings of ducklings with just one or two females in attendance.

Super sites

1. RSPB The Loons and Loch of Banks
2. RSPB Udale Bay
3. RSPB Lough Foyle
4. RSPB Lower Lough Erne
5. RSPB Hodbarrow
6. RSPB Morecambe Bay – Hest Bank
7. RSPB Exe Estuary

WATCHING TIPS

Visiting one of the traditional wintering sites in mid-winter is the easiest way to watch good numbers of Red-breasted Mergansers. They may leave the water and roost on the shoreline, allowing good opportunities for study, but keep at a good distance to avoid disturbance. Communal courtship displays are most common once the birds have reached their breeding grounds, but may be seen among wintering flocks. Young birds form separate flocks at the end of the breeding season, following on from the amalgamation of broods into crèches when the ducklings are still very young, in areas of higher breeding density.

Goosander

Mergus merganser

The Goosander is our largest sawbilled duck. It first nested in Scotland in the late 19th century and now breeds in Wales and down the west side of England but remains absent from Ireland. In winter it is more widespread.

Length 58–66cm

1	2	3	4	5

J F M A M J J A S O N D

How to find

■ **Timing** In areas where it breeds, the Goosander can be seen year-round. In eastern and most of southern England it is a winter visitor, arriving from October and departing from March, though it is most likely to visit the far south-east in mid-winter.

■ **Habitat** Its breeding habitat includes quite fast-flowing rivers as well as upland lakes with plenty of clear open water. It nests in tree holes or rock crevices, so needs access to these close to suitable fishing waters to breed. In winter, migrants from the continent visit lowland lakes and reservoirs, though the breeding birds only make short migrations to lower-lying or more sheltered waters.

■ **Search tips** Visit fast-flowing rivers in the north and west for a chance to see Goosander females with their ducklings – they may be seen on suitable rivers in town centres as well as wilder places. In winter, pairs and small parties visit mostly large, quiet lakes and reservoirs with islands and well-vegetated banks, and you will often find them resting on the shore half-hidden on or among fallen branches and other waterside debris. The long white lower bodies of both sexes are eye-catching from a distance.

WATCHING TIPS

The recent spread of the Goosander's breeding population has enabled many more people to enjoy watching females with their ducklings. Sometimes two or more broods are combined in a crèche, with one female marshalling 15 or more of the charmingly marked ducklings. In winter, Goosanders spend much time resting up, but one interesting behaviour to look for is cooperative fishing, where several birds swim in a line to drive fish to areas where they may be caught more easily. Courtship displays, including communal courtship involving multiple males, may be seen in late winter.

Red Grouse

Lagopus lagopus

Our commonest grouse, the Red Grouse is also a subspecies unique to the UK, distinctly different to its continental counterpart, the Willow Grouse. It is found across much of Scotland, and also north and south-west England, Wales and patchily in Ireland.

Length 37–42cm

1	2	3	4	5

J F M A M J J A S O N D

How to find

■ **Timing** The Red Grouse remains on its moorland habitat year-round, one of the few moorland birds to do so, though it may move to lower altitudes in the depths of winter.

■ **Habitat** Open heather moorland is the only habitat where this species is likely to be seen. Much of this habitat is on high ground, with snow cover in the coldest months. In winter it may visit farmland adjoining the moors, but it avoids anywhere with tree cover.

■ **Search tips** When walking along footpaths through suitable habitat, your most likely view of a Red Grouse will be if you disturb one near the path, whereupon it will spring up from the heather and fly rapidly away on whirring wings, staying low. Patient scanning of the moor may reveal birds some distance away, though they will be hidden among the heather for much of the time. Try searching from a car, driving slowly down lanes through the moorland, and stopping at elevated points to scan. Be patient and wait for birds to move into view. Snow cover makes the birds easier to spot.

Super sites

1. RSPB Forsinard Flows
2. RSPB Corrimony
3. North York Moors
4. Peak District

WATCHING TIPS

Red Grouse form pairs in autumn or winter, and the males defend their territory by calling (especially at dawn) from a somewhat prominent perch, continuing in this role into the following spring while the female is incubating. The song is a curious bubbling rattle, with a croaking or yelping quality. In territorial disputes and courtship, the male displays with tail fanned. Unpaired birds form small flocks in winter, and with luck and patience you may be able to watch them feeding. They are easiest to watch when there is snow on the ground, and you may see them digging out roosting hollows in the snow.

Ptarmigan

Lagopus mutus

This is a true mountain bird, and as such has a very restricted distribution in the UK. Only the highest mountains in Scotland are suitable for it, though it descends a couple of hundred metres down the slopes during the winter months.

Length 34–36cm

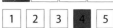

| 1 | 2 | 3 | **4** | 5 |

| J | F | M | A | M | J | J | A | S | O | N | D |

How to find

■ **Timing** The best season to see Ptarmigans depends partly on how far you want to walk – in summer you may need to climb 3,000 metres or more to reach its habitat, but in winter it may be found lower down. However, walking conditions in the mountains are much safer in summer.

■ **Habitat** The Ptarmigan is comfortable among the very open, exposed and rocky terrain that characterises mountainous slopes and summits, where it feeds on the sparse plant matter of various kinds. It tolerates snow cover through much of the year, being able to dig in the snow to reach food, but does try to avoid the snowiest areas in winter.

■ **Search tips** Ptarmigans are well camouflaged at all times of year, moulting into a pure white winter plumage, and so are quite inconspicuous when feeding on the ground. When walking in the mountains, scan slowly, carefully and thoroughly. It sometimes relies on its camouflage for protection when approached and may only flush at the last minute – it may also choose a prominent perch on a hummock or boulder as a lookout post.

Super sites

1. Belach-na-Bah near Applecross
2. Carn Ban Mor
3. Cairngorms

WATCHING TIPS

The Ptarmigan is gregarious so if one bird is found there are probably others nearby – in all-male flocks and family groups of a female plus her young from summer into autumn, then in larger groups through the winter. The birds dig hollows in the snow for roosting, even doing this in summer where most of the ground is snow-free. You may see courtship displays within the groups in winter, the males performing 'dances' with the tail raised and fanned: the black tail is the only striking feature against a snowy backdrop. Males become territorial in spring once pair bonds are formed, guarding their patch from a vantage point. The song is a peculiar rolling frog-like croak.

Tetrao tetrix

Black Grouse

This very striking but shy and elusive grouse has declined drastically in the UK since the mid-20th century, but careful management has helped it to begin a recovery in some areas. It is found in Scotland, northern England and north Wales.

Length 40–59cm

| 1 | 2 | 3 | **4** | 5 |

J F M A M J J A S O N D

How to find

■ **Timing** The Black Grouse is a sedentary species, and can be seen in the same areas throughout the year. Its leks, where males display communally to the females, are most active on early spring mornings, though lekking sometimes takes place at other times.

■ **Habitat** This is a bird of boggy moorland with some trees. It also uses marginal habitats, such as the edges of conifer woodland, farmland or forestry plantations, wet heathland adjoining grassland and similar, but the presence of some trees is essential, as well as areas of bare ground for lekking. It is rarely found close to human habitation.

■ **Search tips** When walking in suitable habitat, there is always the chance that you will flush a Black Grouse, which will fly away in typical low, whirring grouse fashion. When looking for the species, move slowly and quietly and scan well ahead, checking horizontal tree branches as well as the ground. The easiest and probably best way to see the species is to join an organised lek visit, where you will be able to see the birds at their best with no risk of disturbance.

Super sites

1. RSPB Corrimony
2. RSPB Loch Garten
3. RSPB Geltsdale

WATCHING TIPS

Visiting a Black Grouse lek is an unforgettable experience. The assembled males display, fight and leap about, striking dramatic postures to exhibit their plumage to best effect for the benefit of the much drabber females. They call during the display – various purring and harsh growling notes – and there is much rustling of plumage. Organised Black Grouse 'safaris', such as that organised by the RSPB at Corrimony, give you the chance to watch the drama from the shelter of a minibus. Black Grouse are very vulnerable to disturbance, so if you chance upon them while walking (whether they are lekking or not) keep still and watch from a good distance.

Capercaillie

Tetrao urogallus

This huge, turkey-like grouse is a very rare breeding bird in the UK, restricted to pine forests in the central Scottish Highlands. It became extinct in Scotland in the 18th century, but was reintroduced in the early 20th century.

Length 60–87cm

1	2	3	4	5

J F M A M J J A S O N D

How to find

■ **Timing** The Capercaillie is present on its breeding grounds throughout the year, but is never easy to see. Spring is the likeliest season to find one, when males are actively displaying to attract females.

■ **Habitat** Most Capercaillies inhabit the few remaining areas of native Caledonian pinewood in the eastern Highlands and Deeside – these forests are dominated by Scots pine and have a rich understorey of heather and bilberry. Capercaillies may also use commercially planted pine forest, though this is less suitable for their needs.

■ **Search tips** Because of its rarity and vulnerability to disturbance, the Capercaillie should not be sought out with any intensity, particularly in the breeding season. It is difficult to find at any time, because it is so shy, will see you coming from a great distance and can melt away unnoticed into the forest with surprising ease, despite its size. When walking in suitable forest, step softly and slowly, and pause often to scan well into the wood, remembering to check in the trees as well as on the ground. The RSPB's 'Caper Watch' events in Abernethy Forest provide a great chance of disturbance-free sightings.

Super sites

1. RSPB Abernethy Forest
2. Forests in Deeside

WATCHING TIPS

The chances of actually watching Capercaillies for any length of time are slim unless you are watching from a hide. The RSPB's 'Caper Watch', held at the Osprey Centre in Abernethy Forest, runs from the start of April until midway through May. Visitors arrive before dawn to be present in the hide at first light, when lekking male Capercaillies may be on view, performing their extraordinary courtship display. The scheme has a high success rate and, most importantly, minimises the risk of disturbance to the birds. Very occasionally, overly hormonal male 'Capers' lose fear of humans and will actually approach and even attack people!

Red-legged Partridge

This handsome partridge was introduced to the UK from mainland Europe in the late 18th century, for shooting. It has spread to most of England, especially the east side, and up into eastern Scotland. More are released for shooting each year.

Length 32–34cm

1	2	3	4	5

J F M A M J J A S O N D

How to find

■ **Timing** This partridge can be found in the same places at any time of year, rarely wandering more than a kilometre or so from where it was hatched.

■ **Habitat** Found in open countryside, including arable farmland, pasture, dry heathland and the edges of woodland. Areas of bare ground and an open aspect are its key requirements, and on farmland it will prosper best in areas with hedgerows and set-aside land, which will encourage plenty of insects for the chicks to eat.

■ **Search tips** Red-legged Partridges often forage in small parties rather inconspicuously in crop fields, half obscured among the growing plants. Look for their humped dark shapes moving slowly among the crops – at a glance they could be mistaken for Woodpigeons or even rabbits. Also scan the field margins, and along ridges or fences. It is common to encounter groups of Red-legs (or, in late spring and summer, an adult with a brood of chicks) on quiet country lanes through farmland, when they will run rather than fly away from danger.

WATCHING TIPS

Using your car as a hide is a good way to enjoy prolonged views of Red-legged Partridges – they are likely to be alarmed by the outline of a human. Males may be seen calling and displaying to females and rivals in late winter. Often, the female of a pair leaves her first clutch in her mate's care and immediately has a second clutch elsewhere, which she tends, meaning that broods of tiny youngsters are suddenly very common. The chicks follow their parent closely and you may see them being 'shown' food, or being brooded under the adults' wings.

THE BREEDING SEASON

For most birdwatchers, simply seeing a species is only half of the enjoyment. Much more deeply rewarding is to really watch birds, study their behaviour, and gain some insight into how they lead their lives. No aspect of bird behaviour is more fascinating than breeding behaviour – from the sometimes extraordinary courtship displays that set things in motion to the moment a young bird takes to the air for the first time. However, birds are particularly vulnerable during this process, and birdwatchers need to know how to ensure that they do not cause any problems.

Reading behaviour

Understanding bird behaviour is key to making the most of your sightings, and being able to anticipate when something interesting is about to happen. It is also very important for you to know when a bird has stopped behaving 'naturally' and is reacting to your presence, especially in the breeding season.

Once nesting has begun, birds are reluctant to leave the area where the nest is being built. As the season progresses, they invest more and more time and energy in their efforts to successfully breed, and so will become increasingly determined not to be driven away. If you find adult birds allowing you to approach them much more closely than would be 'normal', that may mean you have wandered close to an active nest. Look out for signs of agitation – frequent alarm-calling, raising of the crown feathers, flying repeatedly over your head and sometimes even performing a 'distraction display' where the bird feigns injury to try to lure you away from the nest. Songbirds seen carrying food are always on their way to feed chicks, and if they remain in one spot and watch you for a long time, that could mean that they are reluctant to visit the nest while you are as close as you are.

Birds nesting in town environments can be much more relaxed about human presence. You will probably be able to watch Moorhens and Mallards tending their chicks on your local park lake at point-blank range without any problems. Colonial seabirds can also be very tolerant of human company, but you should still keep a respectful distance at all times, and certainly never be tempted to try to touch a young bird. Your garden birds may be very confiding, but it is still wise not to go near a nest, as predators may be watching.

Wren

Robin

Young birds in trouble

Every year, hundreds of baby birds are needlessly 'rescued' by well-meaning but ill-informed people. Many of these are songbirds that have just fledged, and are unable to fly strongly, if at all. They seem helpless, and of course they are very vulnerable for those first few days out of the nest, but nine times out of ten the best thing you can do to help them is to give them a wide berth.

Only intervene if you can see the bird is in clear, immediate danger – if it is on a road or busy pavement, for example – or if it is obviously injured. If you have to move it, try to place it above ground but as close as possible to where you found it. Its parents will respond to its calls and continue to care for it. Only if the bird is injured should you take it into care. Find a local wildlife rehabilitator who will care for it – they will have the facilities to give it appropriate treatment and manage its successful release.

DANGER ZONE

Most nesting birds will object to people nearby, but some have a more proactive way of dealing with the perceived threat than others. Some birds of prey and seabirds will not hesitate to attack you, and may even do real harm – Tawny Owls and Buzzards have both caused serious injuries to people who were (sometimes unwittingly) too close to their nests. Arctic Terns and Great Skuas are well known for dive-bombing people who walk by their breeding colonies, and Mute Swans can administer a serious beating if they feel you are a danger to their cygnets.

Arctic Tern

Grey Partridge

Perdix perdix

Our native partridge is one of several farmland birds to have undergone severe declines since the second half of the 20th century, although it does still have scattered populations across most low-lying parts of the UK.

Length 29–31cm

| 1 | 2 | 3 | 4 | 5 |

J F M A M J J A S O N D

How to find

■ **Timing** The Grey Partridge can be found in the same general areas year-round, though individuals may move short distances between summer and winter.

■ **Habitat** Like the Red-legged Partridge, this is a bird of open countryside, though it prefers slightly more vegetated habitats in general. Crop fields, pasture or heath with vegetation not much taller than the bird itself is ideal, with safe nesting cover nearby in the form of hedgerows, scrub or woodland fringes. It tends to occur in slightly wetter habitats than the Red-legged Partridge, though both enjoy dust-bathing. The two species do not flock together.

■ **Search tips** Grey Partridges are generally less easy to see than Red-legs, but you should find that similar search techniques are effective. Pay more attention to field margins than centres, and re-scan areas in case birds are moving through taller vegetation. When disturbed, Grey Partridges are more likely to fly than are Red-legs, taking off quite easily and flying away quickly in the typical whirring game bird style. Even half-grown chicks are able to fly. If you disturb a group, the birds will scatter in different directions.

WATCHING TIPS

These game birds are shy and rather difficult to watch. Using your car or some other form of hide is a good bet, but even so the birds may be out of view among vegetation for much of the time. With patience you could enjoy views of the complex interactions within flocks during the winter, with much displaying and aggression between both sexes to establish a dominance hierarchy. In due course, pair bonds become established, and the pair remain together and with their chicks, forming a family group known as a 'covey' which lasts into the autumn. The very steep decline in its population since the late 20th century means that birdwatchers should keep track of local numbers.

Quail

This is our smallest game bird, and the only migratory one. Because it is so elusive and uses different sites year on year, determining its distribution and numbers is difficult, but it may turn up in low-lying farmland almost anywhere in the UK.

Length 16–18cm

| 1 | 2 | 3 | 4 | 5 |

J F M A M J J A S O N D

How to find

■ **Timing** The Quail is one of our latest summer visitors, arriving in late April or early May. It departs in late summer.

■ **Habitat** It seeks out fields of quite high growth – about 1 metre – with hay meadows and cereal fields the most likely places to be used in the UK. It sometimes seeks out bare ground for dust-bathing. It is rather nomadic and unlikely to use the same area year on year – numbers reaching the UK are also highly variable.

■ **Search tips** This is one of the most difficult British birds to see, being extremely secretive and spending most of its time within dense vegetation that conceals it completely. Males holding territory give away their presence with a very distinctive repeated three-syllable call, but hearing the call is the easy part. To see the bird itself will require lots of patience, some very good luck, or both. Positioning yourself where you can clearly see any open patches or tramlines in the field is a good start, and with luck the bird will move across the space. Occasionally stray migrant Quails will turn up in unexpected places, but beware confusion with escaped Japanese Quails (*Coturnix japonica*).

Super sites

1. Berkshire Downs
2. Martin Down
3. South Downs

WATCHING TIPS

A brief glimpse will probably be all the reward you get from a long vigil waiting for a calling male to show itself, especially as the bird has ventriloquism skills and is probably not where you think it is. With great good luck you might witness behaviour such as dust-bathing in a dry hollow, or fly catching with short aerial jumps. Males call to proclaim their territory and to attract females; the calling may continue through the night. In good years densities of calling males may be quite high, with three per square kilometre. Migrants on coasts or islands may be easier to see, if there is less ground cover.

Pheasant
Phasianus colchicus

This unmistakable game bird, introduced to the UK in the 11th century, is common everywhere except the far north-west of Scotland.

Length 53–89cm

| 1 | 2 | 3 | 4 | 5 |

J F M A M J J A S O N D

How to find it

■ **Timing** Pheasants are easy to see at any time of year. Males display and call most in early spring.

■ **Habitat** This species is found in most well-vegetated habitats, with mixed farmland and woodland probably best. It is a frequent visitor to rural and even suburban gardens.

■ **Search tips** In the wider countryside Pheasants are often shy, though birds in parkland and those newly released for shooting can be absurdly tame. Scanning fields and woodland edges from a distance is a good way to find birds engaged in natural behaviour.

WATCHING TIPS

If your garden adjoins woodland or farmland you may attract Pheasants by offering grain or chicken pellets on the ground. Interesting behaviour to look out for includes the male's dramatic 'crowing' display, combined with noisy wing-shaking.

Lady Amherst's Pheasant
Chrysolophus amherstiae

This spectacular East Asian pheasant has been introduced to several parts of the UK, but is currently close to extinction here.

Length 60–120cm

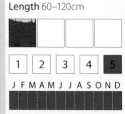

| 1 | 2 | 3 | 4 | 5 |

J F M A M J J A S O N D

How to find it

■ **Timing** Lady Amherst's Pheasant may be seen at any time of year.

■ **Habitat** The native habitat is mountainous forest, but in the UK it lives in dense lowland woodlands, both deciduous and coniferous. It requires a dense understorey of brambles, rhododendrons or similar vegetation.

■ **Search tips** This pheasant is very shy and surprisingly difficult to spot, despite the male's dazzling plumage. Waiting quietly out of view in the very early morning, at intersections within a woodland known to hold the species, will probably give you your best chance to see it.

WATCHING TIPS

The stunning Lady Amherst's Pheasant may soon disappear as a breeding bird in the UK. If you use natural cover or a hide and are prepared to wait, you may see displaying males early in the year, fanning out their remarkable neck ruffs to entice females.

Chrysolophus pictus

Golden Pheasant

A popular aviary bird because of the male's stunning plumage, the Golden Pheasant has become established in a few parts of the UK as a naturalised feral bird. Like the related Lady Amherst's Pheasant, its numbers have been falling steadily.

Length 60–115cm

| 1 | 2 | 3 | 4 | **5** |

J F M A M J J A S O N D

How to find

■ **Timing** Golden Pheasants remain on their breeding grounds all year round. Early mornings and dusk offer the best chance of seeing them, and they are probably easier to see in spring than other seasons.

■ **Habitat** Though a bird of unforested uplands in its native China, the Golden Pheasant lives in low-lying woodland in the UK, principally fairly well-grown conifer plantations but also in mixed woodland with dense ground cover.

■ **Search tips** Like Lady Amherst's Pheasant, this is a very shy and skulking bird. Many encounters are by chance – a bird dashing across the path as you walk down a woodland ride. To maximise your chances, visit one of the key sites either very early in the morning or late in the afternoon and either wait in a well-hidden spot with good visibility through the wood or along paths, or walk around very slowly and quietly, carefully scanning the path ahead of you. In spring you may detect a displaying male by its distinctive crowing call, a very grating and quite shrill double note.

Super sites

1. Sculthorpe Moor
2. Brownsea Island NT

WATCHING TIPS

To watch this shy bird you will need a good hiding place (or the ability to stand very still) and plenty of patience. Golden Pheasants roost in trees but spend their days walking the forest floor, running from danger and only taking flight as a last resort. Outside of captivity they are little studied, so any prolonged observations you manage to make could be of value. Males proclaim and fiercely defend their territories in spring, and court females with a dramatic display, with postures intended to show off the extraordinary plumage. Golden Pheasants are popular aviary birds, and escapees may be seen, including individuals exhibiting unusual colour mutations, and hybrids with other ornamental pheasants.

Red-throated Diver

Gavia stellata

Our smallest diver has a scattered distribution, breeding on small and large water bodies in the north-west Scottish Highlands, Orkney, Shetland, western Northern Ireland and the far north-west of Ireland. In winter it may be found offshore almost anywhere.

Length 53–69cm

1	2	3	4	5

J F M A M J J A S O N D

How to find

■ **Timing** Between April and September to October, the Red-throated Diver will be on its breeding grounds. Non-breeding birds will be at sea year-round, their numbers boosted by breeding birds between late autumn and early spring.

■ **Habitat** Almost any water bodies within the breeding range may be occupied by a nesting pair, though they remain very mobile throughout the nesting season. In particular, those that choose very small pools will constantly commute to nearby larger water bodies to find sufficient fish. They will also visit sheltered bays to forage. In winter, they are usually at sea, although fairly close inshore and in shallow water. May also visit coastal lakes, and occasionally strays to inland waters.

■ **Search tips** Check any lakes in the breeding range, especially those that are undisturbed, bearing in mind that diving birds will be under water for long spells. Also listen for the loud cooing or wailing calls of pairs, and a goose-like cackling given in flight. In winter, scan the sea from an elevated vantage point, looking for an elongated, mostly grey bird, often seen flying very close to the water.

Super sites

1. RSPB Fetlar
2. RSPB Trumland
3. RSPB Hobbister
4. Moray Firth
5. Cardigan Bay
6. RSPB Minsmere

WATCHING TIPS

Red-throated Divers must not be disturbed on their breeding grounds, but it is often possible to safely watch birds on larger lakes from the shore, ideally from hides on nature reserves. The pairs perform mutual displays, and in some areas territorial conflict can be observed between 'neighbours'. By early summer there should be chicks around, hitching rides on their parents' backs. A winter seawatching session may produce large numbers of Red-throated Divers, resting on the sea or flying between good feeding spots.

Black-throated Diver

This diver breeds in north and north-west Scotland but is much scarcer than the Red-throated Diver – it is also absent from Orkney and Shetland. In winter it may be found offshore, especially off north-west Scotland and also Cornwall, and occasionally inland.

Length 58–73cm
Wingspan 110cm

1	2	**3**	4	5

J F M A M J J A S O N D

How to find

■ **Timing** Black-throated Divers generally arrive on their breeding lochs in April or May and depart in September. Outside of this, they can be found offshore, with numbers peaking in mid-winter when some continental birds move westwards to Britain.

■ **Habitat** Less versatile than the Red-throated, this diver requires large, deep freshwater lochs with little or no disturbance. The best sites have small islands offering safe nesting sites – pairs may also use purpose-built rafts – as well as an abundance of fish prey. Particularly large and productive lakes may hold several pairs. In winter, favours sheltered, shallow waters and tends to remain fairly close inshore.

■ **Search tips** Because of the type of waters favoured by breeding pairs, finding them can be quite a challenge. Patient scanning from the shore is necessary, remembering birds will dive for long spells and disappear behind islands. Breeding pairs may give themselves away by their long, drawn-out wailing and yodelling calls. In winter, scan the sea from an elevated spot for flying or swimming birds, and take care to avoid confusion with Red-throated Divers, which will be much more likely away from the few key Black-throated sites. It is always worth checking large coastal lakes and large reservoirs for stray Black-throated Divers in winter.

Super sites

1. RSPB Inversnaid
2. RSPB Forsinard Flows
3. RSPB Loch of Strathbeg
4. Fal Estuary
5. St Ives Bay

WATCHING TIPS

As it is a very rare breeding bird, great care must be taken not to disturb nesting pairs. In practice, nesting sites are often well away from the loch shoreline, and you will need good optics to watch the birds going about their business. As with Red-throated Divers, their behaviour with their chicks is especially tender and engaging.

Great Northern Diver

Gavia immer

This species is our largest diver. It does not breed in Britain, but some non-breeders do oversummer here. In winter numbers increase greatly, and it may be found offshore in most areas, with highest numbers around north-west Scotland and Cornwall.

Length 69–91cm

| 1 | 2 | **3** | 4 | 5 |

J F M A M J J A S O N D

How to find

■ **Timing** Wintering Great Northern Divers arrive from August to September, and stay until early May. Mid-winter sees the highest numbers, but even in mid-summer a few will remain around the north-west coast of Scotland.

■ **Habitat** Like the other divers, it prefers relatively shallow and sheltered seas, and although it tends to be further out than the other two, summering birds in particular may spend much time feeding close inshore. It sometimes strays inland in winter, more often than other divers, and may take up residence on a suitable lake or reservoir for several days or even weeks.

■ **Search tips** Finding Great Northern Divers on a seawatch can be difficult, due to their preference for rather deeper water than other diver species, but bad weather at sea may push them closer inshore. Regular checking of local reservoirs may pay dividends, especially following storms at sea that force them to seek more sheltered places. On the north-west coast of Scotland in spring and summer, check sea lochs, bays and inlets – any reasonably shallow and sheltered water could host a Great Northern Diver.

Super sites

1. RSPB Onziebust
2. RSPB Hobbister
3. RSPB Coll
4. Loch Tarbert
5. Newlyn Harbour
6. Fal Estuary

WATCHING TIPS

The Great Northern Diver in summer plumage is a stunning bird, and visiting the right kinds of coastlines in spring and summer should produce good close views. The best opportunity to watch a winter Great Northern Diver will probably involve an inland bird; these can be surprisingly unconcerned by human observers and give great views. Winter seawatches rarely give close or prolonged views but may give the opportunity to compare all three diver species on the same day.

Little Grebe

Not only is this our smallest grebe, but it is the smallest swimming bird likely to be encountered in Britain. It is very widespread on still or slow-flowing waters and is quick to take advantage of newly created wetland habitats.

Length 25–29cm

| 1 | 2 | 3 | 4 | 5 |

J F M A M J J A S O N D

How to find

■ **Timing** Little Grebes mostly remain on or near their breeding grounds year-round, unless a freeze forces them to move away to find open water.

■ **Habitat** This species breeds on well-vegetated pools and lakes, and slow rivers with plenty of marginal vegetation. The water body needs to have suitable marginal vegetation in which to conceal a nest and contain plenty of prey in the form of small fish, crustaceans and other small underwater life. Once these criteria are met, Little Grebes are at home just as much in town park lakes as they are in upland corries, and for such apparently reluctant fliers are efficient colonisers of new gravel workings and such-like. In winter, may be found on the sea in shallow and well-sheltered bays.

■ **Search tips** Its small size makes this species easy to overlook. Check along the edges of reedy islands and, as with related species, keep in mind that when feeding actively it spends much more time under water than on the surface. In the breeding season especially, single birds and pairs frequently give a distinctive loud and shrill trilling call.

WATCHING TIPS

A Little Grebe in feeding mode will generally work its way along quite close to the shoreline or island edges, diving frequently. You may enjoy the spectacle of it wrestling with prey that seems unmanageably big and lively, or carefully feeding smaller items to its striped chicks. Like other grebes, it often carries its chicks on its back. It is usually double-brooded, the first brood hatching in May. In late autumn, adult and moulting juvenile Little Grebes may gather in small flocks. As winter approaches they often spend more time out over deeper water, giving more prolonged views.

Great Crested Grebe

Podiceps cristatus

Once hunted almost to extinction for its plumage, this beautiful grebe is now common and widespread, though is absent from northern Scotland and some parts of Ireland. It breeds and winters on fresh water but very large numbers also winter offshore.

Length 46–51cm

| 1 | 2 | 3 | 4 | 5 |

J F M A M J J A S O N D

How to find

■ **Timing** Breeding pairs of Great Crested Grebes are on territory year-round, beginning their nesting activity as early as January. Winter gatherings, both at sea and inland, peak in December or January.

■ **Habitat** This grebe nests on lakes, reservoirs, canals and slow rivers, with good supplies of fish. The floating nest is usually anchored to waterside vegetation but may be attached to a bare island or even a man-made structure, meaning the species may nest on less well-vegetated waters. It readily breeds on town park lakes, where it will become as fearless of people as the park ducks and swans. After the breeding season, young birds and some adults may congregate on larger waters, including sheltered seas.

■ **Search tips** The Great Crested Grebe stands out as a slim, mostly pale figure among the chunkier and darker shapes of other common lake waterfowl. When feeding it dives often, but spends much time resting on the water well away from the shoreline. Winter seawatches, especially off more southern coastlines over sheltered water, may reveal large rafts of Great Crested Grebes, although the species is not very gregarious at other times.

WATCHING TIPS

Most birdwatchers are especially keen to witness the famous courtship dance of the Great Crested Grebe. Established pairs may 'perform' at any time of year but the full display is most frequent at the start of the breeding season, in late winter. Watch out for pairs swimming in close proximity – the display may begin with subtle head-shaking before progressing to the dramatic 'weed dance'. Later in the season, the pair may be observed tending their chicks, which follow their parents around with incessant loud calls. Town parks offer good opportunities to watch feeding behaviour close up.

Podiceps grisegena

Red-necked Grebe

This species is the scarcest of our five regularly occurring grebes. A handful may oversummer on freshwater lakes, and breeding has been recorded. A couple of hundred visit in winter, mainly at sea along the southern and eastern coasts.

Length 40–50cm

| 1 | 2 | 3 | 4 | 5 |

J F M A M J J A S O N D

How to find

■ **Timing** Wintering Red-necked Grebes are present between October and March, with the highest numbers in mid-winter. Severe weather further east forces more birds to head our way. Oversummering birds are likely to occupy the same site for weeks, but because of the risk of disturbance of birds that may be breeding, localities are not always made public.

■ **Habitat** In winter, Red-necked Grebes seek out sheltered and shallow coastal waters – bays, firths and sea lochs. Passage migrants may visit fresh water close to the coast, while summering birds may venture further inland, favouring large and undisturbed lakes and reservoirs.

■ **Search tips** Seawatching from the south or east coast in winter gives you the best chance of seeing this species, especially following bad weather on the continent. Choose places that give you a good view over an area of reasonably sheltered water, and scan patiently – feeding birds will be under water much of the time. Storms at sea may force Red-necked Grebes to take shelter on estuaries or fresh water near the coast. It is always worth keeping an eye open for passage migrants or oversummering birds when visiting suitable lakes.

Super sites

1. Firth of Forth
2. Filey Bay
3. Sheringham
4. RSPB Dungeness

WATCHING TIPS

With Red-necked Grebes in winter, beware confusion with Great Crested Grebe, which is present at sea in large numbers in many locations and in winter plumage looks similar. The Red-necked Grebe's dusky grey neck is probably the best identification feature when watching from a distance, but with more familiarity, other subtle differences become more apparent. Although there are far fewer birds inland than at sea, if you manage to find one you should enjoy much better views than when seawatching, and possibly the chance for side-by-side comparisons with Great Crested Grebe.

Slavonian Grebe

Podiceps auritus

This beautiful grebe is a very rare breeding bird in Britain, with some 40 pairs in the Scottish Highlands. In winter, a few hundred birds from further east visit much of our coastline, and also coastal and inland fresh water.

Length 31–38cm

| 1 | 2 | 3 | 4 | 5 |

J F M A M J J A S O N D

How to find

■ **Timing** The best time to see breeding birds is between mid-April and July. Views at RSPB Loch Ruthven are somewhat better in the mornings, due to the direction of light relative to the hide's position. Wintering birds are best looked for between October and March, with mid-winter best.

■ **Habitat** Breeding birds require lush, shallow lakes with emergent vegetation, and they may use quite small pools if there is sufficient prey. In winter, sheltered and shallow seas, including bays and estuary mouths, hold most birds, with some opting for coastal lakes and reservoirs (especially following storms at sea).

■ **Search tips** RSPB Loch Ruthven has an established population of Slavonian Grebes, and this reserve offers the best and safest opportunity to watch breeding birds. Views should be had easily from the hide there – few other water birds use the loch so there is little risk of confusion. In winter, although more birds are at sea, better views can often be had by visiting suitable lakes or reservoirs along the coast. For winter seawatching, pick fairly calm days and quiet, sheltered patches of sea to survey. Powerful optics may be needed to find these very small birds at sea.

Super sites

1. RSPB Loch Ruthven
2. Largo Bay
3. Gosford Bay
4. Bedmennarch Bay
5. RSPB Titchwell
6. Portland Harbour
7. Poole Harbour
8. Pagham Harbour
9. RSPB Dungeness

WATCHING TIPS

Slavonian Grebes can be watched at fairly close range at RSPB Loch Ruthven. Early in the season you may witness the mutual courtship display, while later on the adults will be escorting their chicks around the sedge beds. In winter, birds sometimes come quite close inshore, giving good views, but you are more likely to have prolonged and closer views at coastal lakes. Slavonian Grebes are wary birds so use hides when possible and wait – a feeding bird will work its way across a wide area in a short space of time.

Black-necked Grebe

The Black-necked Grebe maintains a small breeding population, mostly well inland. It is uncommon in winter too, with 100 or so present mainly at southern sites on the coast. It is also found inland more often than the other two rare grebes.

Length 28–34cm

| 1 | 2 | 3 | **4** | 5 |

J F M A M J J A S O N D

How to find

■ **Timing** Black-necked Grebes are on their breeding grounds between late March and September, and numbers at wintering sites peak in mid-winter. Passage migrants could turn up on almost any good-sized water body in England, Wales and southern Scotland.

■ **Habitat** Breeding sites are well-vegetated shallow and often small lakes and pools, with a rich supply of crustaceans and other prey. Varying water levels from year to year may mean that a site that hosts breeding birds one year may be deserted the next. In winter, deeper and more sparsely vegetated lakes may attract them. At sea, shallow harbours and bays are best, with most UK birds shared between a small number of favoured sites.

■ **Search tips** As a very rare breeding bird, the Black-necked Grebe's breeding sites may not be publicised – the often transient nature of suitable sites means that it is difficult to establish a visitor infrastructure geared towards seeing the species. No such secrecy surrounds the whereabouts of wintering birds, which often remain for several weeks. A patient and thorough scan of open water is necessary as feeding birds make frequent long dives.

Super sites

1. Loch Ryan
2. Lower Derwent Valley
3. Lower Bittel Reservoir
4. Staines Reservoirs
5. Portland Harbour
6. Studland Bay
7. Langstone Harbour
8. RSPB Dungeness

WATCHING TIPS

If your local lake hosts more than one Black-necked Grebe through winter, you see territorial or courtship behaviour in early spring, even if the birds in question do not remain on the lake to breed. Take great care not to disturb any birds showing signs of breeding behaviour. Because this species needs sheltered water, birds at sea may be close inshore and give good views – the same goes for birds wintering on lakes and reservoirs.

Fulmar

Fulmarus glacialis

This is the most common of our 'tubenose' seabirds, breeding on suitable cliffs around most of the coastline of the UK and Ireland. When not breeding it may be seen offshore, and very occasionally storm-driven birds turn up inland.

Length 45–50cm
Wingspan 102–112cm

| 1 | 2 | 3 | 4 | 5 |

J F M A M J J A S O N D

How to find

■ **Timing** Fulmars occupy their breeding cliffs for longer than most other seabirds, settling down as early as February and remaining until August. Between mid-August and October they move further out to sea to moult, but will come closer inshore and regularly visit their cliffs through winter.

■ **Habitat** Steep cliffs of all kinds are used for breeding, including relatively soft and low sandstone cliffs and even coastal buildings in a few places. They are not usually closely colonial, with nests quite well spaced out, even at sites where few other cliff-nesting species are present.

■ **Search tips** At breeding colonies, Fulmars are usually much in evidence as they wheel tirelessly around and skim the cliff edge. Their distinctive flight style, interspersing rapid flaps with long glides on stiff, straight wings, helps distinguish them from the more languidly flapping gulls that will probably also be present. Pairs on the nest usually look gleaming white, and huddle close together on a ledge or in a small crevice, often calling together with a harsh guttural rattle. Fulmars readily follow boats at sea and are attracted by 'chum'.

WATCHING TIPS

Watching Fulmars from the clifftops at a seabird colony is exhilarating – the birds will often go by at very close range, unlike most of the other species which remain at lower levels in the airspace. Take great care not to approach a nest though, for the birds' sake and your own (they may eject their smelly stomach contents at you – a very nasty and effective way of deterring intruders). Most pelagic trips will result in close Fulmar encounters – the birds will glide alongside the boat almost within touching distance.

Calonectris diomedea   # Cory's Shearwater

This bird breeds on Mediterranean coasts and islands, wanderers sometimes pass south-western coastlines in summer or autumn.

L 45–46cm **WS** 105–125cm

1	2	3	4	**5**

J	F	M	A	M	J	J	A	S	O	N	D

How to find it

- **Timing** August is probably the best month to look for this seabird. Strong onshore winds greatly increase the likelihood of a sighting.
- **Habitat** When not breeding, this species ranges widely at sea, often hundreds of miles from land. Many feed in the Bay of Biscay.
- **Search tips** Visit prominent headlands in south-west England or south-west Ireland, under suitable weather conditions, for the best chance of a sighting. Late-summer boat trips towards or across the Bay of Biscay should produce sightings.
- **Super sites** 1. Bridges of Ross, 2. Galley Head, 3. Pendeen, 4. Porthgwarra, 5. Berry Head.

WATCHING TIPS

Watch for the typical tilting or 'shearing' flight on very straight wings, low and often dipping between the wave crests. When watching from a boat, you may see birds on the water as well. Most feeding takes place at night – by day, large 'rafts' may gather to rest.

Puffinus gravis   # Great Shearwater

Another ocean wanderer, this large shearwater is most likely to be seen from headlands that project into the Atlantic.

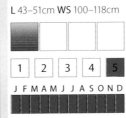

L 43–51cm **WS** 100–118cm

1	2	3	4	**5**

J	F	M	A	M	J	J	A	S	O	N	D

How to find it

- **Timing** As with Cory's Shearwater, this is primarily a late-summer visitor. The vast majority of sightings are between July and September, and August is the best month to look. Onshore winds are necessary.
- **Habitat** These open-sea wanderers range further north than Cory's Shearwater, reaching well north of Scotland.
- **Search tips** The options are either patient seawatching in suitable weather conditions, or a boat trip out into the Atlantic. Look for a powerfully flying shearwater, dark on top with a prominent white collar.
- **Super sites** 1. RSPB Sumburgh Head, 2. Bridges of Ross, 3. Galley Head, 4. Pendeen, 5. Porthgwarra.

WATCHING TIPS

As with Cory's Shearwater, you are likely to get closer and more prolonged views from a boat than on land, with a good chance of watching groups of birds resting on the sea or following the boat. However, if the weather is right birds may pass very close offshore.

Sooty Shearwater

Puffinus griseus

This shearwater breeds close to the Antarctic and migrates to winter in northern seas, passing our coasts on its return journey.

L 40–51cm WS 94–109cm

1	2	3	4	5

J F M A M J J A S O N D

How to find it

■ **Timing** Sooty Shearwaters pass our coastlines mainly in late summer and autumn, usually heading south. As with other seagoing species, pick a day with strong onshore winds to give the best chance of closer views.

■ **Habitat** The Sooty Shearwater tends to only be seen when strong northerly winds are blowing. The east coast is generally better than the west.

■ **Search tips** Seawatch from suitable headlands. Beware confusion with other dark, long-winged seabirds (juvenile Gannet and dark-morph skuas).

■ **Super sites** 1. Fife Ness, 2. Hartlepool, 3. Flamborough Head, 4. Sheringham, 5. Pendeen, 6. Porthgwarra.

WATCHING TIPS

Birds seen from our shores are likely to be heading south at full speed towards their breeding quarters so views may be brief, but on a good day hundreds may pass the headlands, making for a wonderful spectacle against the backdrop of a tempestuous sea.

Balearic Shearwater

Puffinus mauretanicus

A smallish and rather dark shearwater, this species breeds in the Mediterranean and is considered to be critically endangered.

L 34–38cm WS 83–93cm

1	2	3	4	5

J F M A M J J A S O N D

How to find it

■ **Timing** The migratory movements of this bird are still not well understood. After breeding many disperse north-westwards out of the Mediterranean, making summer and autumn the best time to see it here.

■ **Habitat** This shearwater does not wander the seas as widely as the other species, but it may still be many miles offshore, unless pushed close to land by strong winds.

■ **Search tips** From suitable seawatching points, search for a shearwater with rather drab, uncontrasting brown plumage. Balearic Shearwaters may fly in company with Manx Shearwaters, which look crisply black and white by comparison.

■ **Super sites** 1. Strumble Head, 2. Pendeen, 3. Porthgwarra, 4. Portland Bill.

WATCHING TIPS

This is a rather nondescript species, but with luck you will be able to compare with Manx Shearwater, which will reveal the subtle differences between the two species. The global rarity of Balearic Shearwater means that logging sightings is especially important.

Puffinus puffinus
Manx Shearwater

This is our only breeding shearwater. Most of its world population breeds on scattered coastal and island sites in south-west England, Wales, north and west Scotland and Ireland. When not breeding, it wanders the sea on all sides of the British Isles.

Length 31–36cm
Wingspan 76–88cm

1	2	3	4	5

J F M A M J J A S O N D

How to find

■ **Timing** Most birds are on their breeding grounds between March and July, though adults only come ashore to feed their chicks at night. Migrants pass offshore elsewhere in spring and autumn, and wandering birds may also be seen in winter, though rarely in large numbers. Onshore winds produce more sightings.
■ **Habitat** The Manx Shearwater is a burrow-nester – this plus its nocturnal nest visits help protect it and its chicks from gulls and skuas. It therefore needs soft soil into which it can tunnel (or old rabbit burrows), which adjoin the sea. Islands that are free of mammalian predators (especially rats) hold the most successful colonies. At sea, it roams widely but passes closer inshore on migration and in bad weather.
■ **Search tips** Some nesting colonies, for example on Skomer, can be visited by arrangement, but some have no access. However, shearwaters often gather in rafts on the sea near colonies as dusk approaches. When seawatching, watch for small, long-winged birds that flash alternately black and white as they tilt and 'shear' low over the water. In flapping flight they could be confused with auks, but while auks flap almost continuously, shearwaters intersperse flapping with stiff-winged glides.

Super sites

1. RSPB Balranald
2. RSPB Hoy
3. Western Irish headlands
4. RSPB Bempton Cliffs
5. Flamborough Head
6. RSPB Ramsey Island
7. Skomer
8. Skokholm
9. Portland Bill

WATCHING TIPS

An evening or all-night visit to a breeding colony is an unforgettable experience, with the chance to watch the shearwaters assembling on the water, waiting for darkness to fall. Later, the birds arrive on land under torchlight and (hopefully) a moonlit sky, the adults exchanging a chorus of weirdly demonic calls as they land clumsily and stumble to their nesting burrows. In optimum weather conditions, you may see hundreds of Manx Shearwaters on an autumn seawatch in particular, while boat trips out of almost any British harbour could produce great, eye-level views.

Storm Petrel

Hydrobates pelagicus

Though barely larger than a sparrow, this long-lived bird spends most of its life over the open sea, coming ashore only to breed in its coastal and island colonies, most of which are around north and west Scotland and Wales.

Length 14–18cm
Wingspan 36–39cm

| 1 | 2 | 3 | **4** | 5 |

J F M A M J J A S O N D

How to find

■ **Timing** Storm Petrels are on their breeding grounds between early May and September or October. They come to land only at night, to avoid the predatory attentions of larger seabirds. Post-breeding, they migrate to southern African waters, and may be seen offshore through the autumn, especially when onshore winds force them to travel closer to land.
■ **Habitat** Nesting sites are in various kinds of small crevices, from rabbit burrows to cracks and gaps in the walls of abandoned buildings. Most thriving colonies are on rat-free islands with little or no human disturbance. At sea, ranges widely but comes inshore when driven there by storms, and is attracted to fishing boats.
■ **Search tips** Many nesting colonies are inaccessible. The RSPB's Mousa reserve in Shetland is an exception, with ferry trips offering visitors the chance to see petrels at the Iron Age broch where they nest. Boat trips can also be good for Storm Petrel sightings, especially around the vicinity of breeding colonies in summer. When seawatching for this species, look for a tiny, dark, white-rumped bird rising and falling over the waves with a restless, almost bat-like flight.

Super sites

1. RSPB Mousa
2. Carnsore Point
3. RSPB Ramsey Island
4. Skokholm
5. Isles of Scilly
6. Pendeen

WATCHING TIPS

Visiting a nesting colony such as RSPB Mousa, or Skokholm off Pembrokeshire, offers a special insight into how these remarkable birds lead their lives – opt for an evening or night visit to see (and hear) plenty of activity. The birds have an extraordinary purring song, rather Nightjar-like but with a frantic, almost gasping quality. Playback of song recordings has been used to monitor birds at potential nesting sites on nature reserves. Joining a dedicated 'pelagic' trip where fish offal and oil is thrown out to attract seabirds can produce wonderful close views of Storm Petrels, looking impossibly delicate as they flit and dip, dangling their long legs to touch the water's surface.

Oceanites oceanicus

Wilson's Petrel

The scarcest of the storm petrels to regularly occur in British waters, Wilson's Petrel is mainly seen in late summer, from headlands along the south-west coasts (England, Wales and Ireland), and from pelagic trips into the same waters.

Length 15–19cm
Wingspan 110cm

| 1 | 2 | 3 | 4 | **5** |

J F M A M J J A S O N D

How to find

■ **Timing** Most Wilson's Petrels are seen between mid-July and mid-September. There is little chance of sightings from land unless there are strong winds from the west.
■ **Habitat** Like other petrels this is a pelagic species, wandering at sea (though spending most of its time in inshore waters) and able to tolerate quite rough sea conditions. It is usually seen flying low over the water, especially when actively feeding.
■ **Search tips** The headlands of south-west Ireland and the far west of Cornwall produce the most sightings of this species. On days of suitable wind conditions, arrive early and find a position elevated enough to give you a view between the wave crests. Most of the main seawatching sites have established watchpoints, some even offering shelter from the weather. A telescope will probably be necessary – scan at low magnification and then zoom in when you find a bird. Pelagic trips, especially those out of the Isles of Scilly, go into main feeding areas for Wilson's Petrels and give you the chance of close views in calmer conditions.

Super sites

1. Bridges of Ross
2. Brandon Point
3. Cape Clear
4. Pendeen
5. Porthgwarra

WATCHING TIPS

Wilson's Petrel resembles Storm Petrel quite closely, but with good views you will see the darker underwings. There are also some behavioural differences – while both species will 'patter' their feet on the surface, Wilson's Petrel actually appears to 'walk on water', stretching out its long legs and appearing to hop along or even stand on the surface, using headwinds to keep its position. Join a pelagic trip for your best chance of witnessing this behaviour. Wilson's Petrel is probably one of the most difficult British species to see from land, but views on pelagics can be wonderful – and give you the chance to see other scarce seabirds as well.

Leach's Petrel

Oceanodroma leucorhoa

This is an enigmatic seabird, with a sizeable breeding population on the remote island of St Kilda but only small numbers on the closer islands around northern Scotland. Under very specific conditions, though, hundreds of migrants may be seen.

Length 19–22cm
Wingspan 45–48cm

J F M A M J J A S O N D

How to find

■ **Timing** Because the breeding colonies are difficult or impossible to access, your best chance of seeing a Leach's Petrel is during the autumn migration period, especially in September and October. Gales at sea drive birds inshore and sometimes even inland.

■ **Habitat** Like the Storm Petrel, this species is very vulnerable to predation at the nest, so visits its island colonies by night and nests in burrows, which it may excavate itself in soft ground. When not breeding it may travel and feed miles offshore but storms force it to more sheltered inshore waters.

■ **Search tips** Seawatching or pelagic boat trips offer the best chance to catch up with this species. Windy and stormy weather in mid-autumn can bring hundreds close inshore, where they can be viewed from west coast headlands – they will also sometimes head upriver to escape sea storms, with the Dee and Mersey Estuaries being particularly good places from which to watch. After a very stormy autumn night, it may be worth checking local rivers and reservoirs for Leach's Petrels that have been forced to move well inland.

Super sites

1. RSPB Dee Estuary – Point of Ayr
2. The Wirral coastline
3. Spurn Point
4. Strumble Head
5. Sheringham

WATCHING TIPS

Most autumns produce a few exceptional 'Leach's days' when there are strong north-westerly winds hitting the right stretches of coastline. Views can then be very close, and involve dozens or even hundreds of birds going by, sometimes even lingering to feed in between the waves. The best days do unfortunately involve the most atrocious weather, so dress warmly with a wind-and-waterproof outer layer, and choose a sheltered viewpoint. If taking photographs, use a cover to protect your camera from spray. If you are lucky enough to find a bird on an inland reservoir, you should be able to enjoy lengthy views.

Morus bassanus

Gannet

The Gannet is our largest seabird. It breeds on mainland and island clifftops primarily around Scotland and Ireland, with a few more in northern England and Wales. At other times of year it may be seen offshore almost anywhere.

Length 87–100cm
Wingspan 165–180cm

| 1 | 2 | 3 | 4 | 5 |

J F M A M J J A S O N D

How to find

■ **Timing** Gannets begin to settle on their nesting grounds from February, and the last young birds leave in September, having been abandoned by their parents some weeks before. Birds can be seen offshore in all weathers and at all times of year.
■ **Habitat** This species requires exposed flat clifftops rather than sheer ledges, usually on undisturbed islands. Nesting pairs cover the whole available surface, the nests close together but very evenly spaced, each pair defending the immediate area around from their neighbours. When fishing, it tends to remain fairly close inshore and is not especially prone to wander to very deep water.
■ **Search tips** Gannets are impossible to miss at their colonies, and throughout summer will be commuting along the coastline nearby looking for good fishing spots. In autumn and winter they can be found offshore in most areas, though often too far out to be noticed with the naked eye. From seawatching spots, look out not just for the bright white adults, but for juveniles, which are uniformly dark, and the variegated dark and white subadults. They often fly a good height above the waves, and may follow boats.

Super sites

1. RSPB Troup Head
2. Bass Rock
3. Farne Islands NT
4. RSPB Bempton Cliffs
5. RSPB Grassholm

WATCHING TIPS

Visiting a gannetry is a wonderful experience – you can easily while away a day watching the comings and goings of the adults as they work hard to keep their growing chicks supplied with fish. Mainland colonies can be watched from the clifftops, while there are boat trips to some of the best islands in summer. Gannet watching from a boat is especially exciting, as the birds plunge-dive for fish all around you. Beaches near breeding colonies may also give you good views of birds feeding close inshore, especially at high tide.

SEAWATCHING

On the face of it, seawatching is one of the simplest (and certainly least energetic) forms of birdwatching – just find yourself a comfortable place to sit looking out to sea, and watch. Any point along the coastline of Britain will produce sightings of *some* birds, but to see good numbers and to give yourself a chance of finding the more difficult seabird species, you will have to be a bit more selective in terms of when, where and how you seawatch.

Time of year

Most seabirds come to shore to breed, and spend the rest of their time wandering at sea, often far from land and hundreds of miles from the breeding grounds. Therefore, unless you are visiting a breeding colony, the best times to seawatch are in spring and autumn, when seabirds are travelling to or from their colonies. Exactly which month is best depends on the bird species – for example, Pomarine Skua spring migration is best observed in May, while the main Little Auk passage down the east coast is as late as November. Some grebes, divers and sea ducks overwinter around our shores, and non-breeding individuals can be seen year-round.

Weather

It's a sad fact of life that the best weather conditions for seawatching are generally the least pleasant conditions to be at the seaside. In fine and still weather seabirds will often migrate well offshore, out of range of even the most powerful telescopes. Stormy wet weather, though, with strong onshore winds, pushes the birds closer to the shelter of land. The exact position of your chosen seawatching point determines which wind direction will be most productive – if you choose a north-facing spot, for example, you'll want northerly winds to push the birds your way. Needless to say, sitting still in filthy weather necessitates warm waterproof clothing.

Location

Seabirds will generally take the shortest route along the coastline, passing close to headlands but taking short cuts across bays. Prominent headlands are therefore the best points from which to watch, and there are productive headlands all around the British coastline (see below).

Once you have picked a site, you need to find a spot from which you have a wide and somewhat elevated view. Most seabirds will travel just above the waves, so you'll need to be high enough to see birds moving between the wave crests – high up the beach or on a sea wall are often good. Some sites have seawatching hides, where you can watch in shelter while swapping notes with fellow seawatchers.

How to watch

Most seawatchers use both binoculars and telescopes, the former to scan and the latter for a closer view of the birds you find. For both optics having a large depth of field and a wide angle of view is extremely useful when seawatching, enabling you to view a larger 'chunk' of sea in one go.

Once you find a bird, the next challenge is identification. You may only get a very basic impression of pattern and colours. Because the bird will often be too distant for you to make out finer details of plumage, features such as shape and flying style become especially important.

Arctic Tern chasing Arctic Skua

BOAT TRIPS

As long as you have the stomach for it, a boat trip is a great way to get very close to the more elusive seabirds. Special 'pelagic' trips go out into productive waters, and may throw out 'chum' (a revolting fishy concoction) to attract the birds. Pelagic trips are especially good for seeing shearwaters, petrels and skuas.

Where to watch

Here are some of the best seawatching spots around the British coastline.

- Porthgwarra, Cornwall
- Pendeen, Cornwall
- Portland Head, Dorset
- Selsey Bill, Sussex
- Dungeness, Kent
- Great Yarmouth, Suffolk
- Sheringham, Norfolk
- Spurn Head, Yorkshire
- Flamborough Head, Yorkshire
- Fife Ness, Fife
- Tarbat Ness, Rosshire
- Bowness-on-Solway, Cumbria
- Dee Estuary, Cheshire
- Strumble Head, Pembrokeshire
- Bridges of Ross, Co. Clare
- Mizen Head, Co. Cork

Juvenile Guillemot

Cormorant

Phalacrocorax carbo

Though ostensibly a seabird, the Cormorant also breeds well inland, substituting lakeside trees for sea cliffs. It is common and widespread throughout Britain, even using town parks, especially through the winter months when youngsters and breeding birds disperse.

Length 80–100cm
Wingspan 130–160cm

| 1 | 2 | 3 | 4 | 5 |

J F M A M J J A S O N D

How to find

■ **Timing** Cormorants can be seen offshore and on inland waters year-round. They are concentrated at breeding colonies in spring and summer and more generally distributed at other times.

■ **Habitat** Coastal Cormorants nest on lower cliffs and among rocks on both the mainland and on islands, needing quite deep ledges or sizeable crevices. Inland, they nest colonially in trees adjoining large lakes or reservoirs, building sizeable stick nests. When not breeding they can be seen fishing on the sea quite close inshore, and on good-sized lakes, reservoirs and rivers.

■ **Search tips** Cormorants are eye-catching in flight as they fly low over the sea or much higher over land. On the water, look for a bird swimming so low that its back may be almost submerged, just showing a long neck and uptilted head. In between fishing forays, birds often rest on any handy structures projecting from the water, and their habit of standing with wings spread makes them even more noticeable.

WATCHING TIPS

Large town parks with lakes containing good fish stocks may attract Cormorants, especially in winter. These birds become quite approachable and can be watched at close range as they fish the waters. Courtship and breeding behaviour is probably more easily watched at inland colonies than at sea; also, as birds tend to nest in closer proximity in trees than on cliffs, there are likely to be more interactions to observe. Two subspecies of Cormorant occur in Britain, *P. c. carbo* and *P. c. sinensis*, and with careful study the two can usually be distinguished. Young Cormorants are much paler than adults, and are often mistaken for divers. Check the bill shape and posture on the water if you are unsure.

Phalacrocorax aristotelis

Shag

This smaller relative of the Cormorant is a true seabird, rarely seen inland. A large proportion of its breeding population is concentrated at only a few sites, and it can be seen around most of our coastline all year.

Length 65–80cm
Wingspan 90–105cm

| 1 | 2 | 3 | 4 | 5 |

J F M A M J J A S O N D

How to find

■ **Timing** Shags occupy their nesting sites from late February until late summer, but adults remain close to the breeding sites all year round. Young birds may roam a little further, and birds are occasionally found inland, on rivers, lakes or reservoirs, after sea storms.

■ **Habitat** Shags may nest on precipitous cliffs alongside other seabirds, but will also use lower rocky shorelines and boulder beaches. They need good-sized ledges or hollows, and on tall cliffs are generally found quite low down. They may be seen fishing in sheltered waters close to shore.

■ **Search tips** Scan a likely-looking cliffside carefully, paying particular attention to the lower slopes and to any plateaus or sheltered crevices. Shags often sit on their bellies, making them less obvious at a casual glance than when standing 'tall'. Also scan the sea below for Shags swimming or flying very close to the water, looking very slim and elongated with their long necks and tails. Like Cormorants, Shags need to dry out their open wings after swimming, and may be found resting on offshore rocks near to the breeding sites.

WATCHING TIPS

When visiting a busy seabird colony, it's easy to overlook the Shags among other more active or vocal species. It's worth taking the time to admire their beautiful green- and violet-glossed plumage, and observe their behaviour. Unlike many seabirds they actually construct a nest, from whatever floating seaweed they can find, and some males have two mates and two sets of chicks to provision. On seawatches, you are likely to see both Shags and Cormorants, a good opportunity to learn how to separate the two species at a distance. Although Shags are scarce inland, individuals that do stray from the coast may make long stays on rivers or reservoirs.

Bittern

Botaurus stellaris

This may seem an almost mythical bird, given its rarity and shy, skulking nature. It has responded well to conservation efforts and is slowly increasing, both as a breeding and wintering bird (though most winterers are just visiting from further east).

Length 70–80cm

| 1 | 2 | 3 | 4 | 5 |

J F M A M J J A S O N D

How to find

■ **Timing** Bitterns are easier to see in winter, especially in very cold weather, than summer, but may be seen at any time of year. The remarkable foghorn-like 'booming' call of the males is mainly heard in spring.

■ **Habitat** This is a bird of dense and extensive reedbeds, within which are channels or small pools of open water where it can search for fish and other prey. In winter, it requires less space, and freeze-ups may force it to use quite small reedbeds.

■ **Search tips** Peering into a reedbed is a thankless task, and as this is where Bitterns spend most of their time, views can be few and far between. In summer particularly, you are most likely to see one when it takes flight, moving between feeding grounds – a broad-winged, heavily flapping tawny bird, with long legs but a pulled-in neck. Foraging Bitterns will come to the edge of the reedbed from time to time, and this is when you might see them in action. Most of the best Bittern breeding sites are nature reserves with hides from which views can usually be had, though you may have to wait a while.

Super sites

1. RSPB Leighton Moss
2. RSPB Titchwell
3. RSPB Minsmere
4. RSPB Rye Meads
5. Lee Valley Regional Park
6. Rye Harbour
7. RSPB Dungeness

WATCHING TIPS

Waiting for a cold snap before visiting a site known to host wintering Bitterns is a good strategy for getting good and prolonged views of the species. A partial freeze limits the areas where the birds can feed, and they will often be forced out into the open. For such large birds, they can be surprisingly difficult to spot, and if they sense danger they may freeze in position for long spells. Bitterns may form communal roosts in winter, and arriving birds can be watched flying in at dusk and pitching down into the reeds. A spring visit to a suitable reedbed should allow you to hear the famous 'boom', though sightings are considerably more difficult at this time of year.

Bubulcus ibis

Cattle Egret

Widespread in Europe, this egret is spreading northwards and now a few dozen regularly winter here, mainly in the south and east.

Length 48–53cm

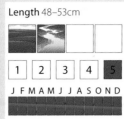

| 1 | 2 | 3 | 4 | 5 |

J F M A M J J A S O N D

How to find it

■ **Timing** Cattle Egrets may be seen at any time of year, but winter is the likeliest season. The first UK breeding was in 2008, following a large winter influx.
■ **Habitat** It often follows cows or horses, looking for insects disturbed by the animals' feet. Low-lying grazed meadows with nearby still or slow-flowing fresh water is ideal feeding ground. It nests in trees, and may join Little Egret colonies.
■ **Search tips** Scan fields and water edges for a small, squat egret, distinguished from the Little Egret by its yellow bill.
■ **Super sites 1.** Somerset Levels, **2.** RSPB Minsmere, **3.** Thorney Island, **4.** RSPB Dungeness.

WATCHING TIPS

The feeding behaviour of these egrets is interesting – they loiter almost at the heels of livestock, and may even hitch a ride on a cow's back. Further breeding in Britain seems likely – check Little Egret colonies, especially if Cattle Egrets have wintered locally.

Ardea alba

Great White Egret

Though it is yet to breed in Britain, this egret is visiting in increasing numbers. Most records are from the south and east.

Length 85–102cm

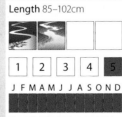

| 1 | 2 | 3 | 4 | 5 |

J F M A M J J A S O N D

How to find it

■ **Timing** Winter and spring are the likeliest times to see Great White Egret, but there are records from all times of the year, and birds often remain at the same site for weeks.
■ **Habitat** This is a bird of fresh and brackish wetlands, usually with extensive areas of water, though it may also fish along small channels in wet pasture.
■ **Search tips** Scan along the edges of lakes and islands, looking for a very tall and slim egret stalking slowly along or standing stock-still. As well as size, the yellow bill and snake-like neck helps distinguish it from the Little Egret.
■ **Super sites 1.** RSPB Leighton Moss, **2.** Rutland Water and other Midlands reservoirs, **3.** RSPB Strumpshaw Fen, **4.** Somerset Levels, **5.** RSPB Dungeness.

WATCHING TIPS

This egret is rather like the Grey Heron in its habits, tending to move in slow motion when hunting. This makes for good watching opportunities. Most places that attract the species will also have Little Egrets, allowing for direct side-by-side comparisons.

Little Egret

Egretta garzetta

Since its first breeding in Britain in 1996, the Little Egret has spread fast and now more than 150 pairs breed, mainly in the south and east. By winter more than 1,500 are here, again mainly in the south and east.

Length 55–65cm

| 1 | 2 | 3 | 4 | 5 |

J F M A M J J A S O N D

How to find

■ **Timing** Numbers start to build up in late summer, but provided you visit the right sites this egret is easy to find throughout the year. The chances of finding one away from the main areas is highest in winter.

■ **Habitat** Little Egrets mainly hunt in or by shallow water. This can be fresh, brackish or salt, and they forage anywhere from tiny temporary rain pools in pasture to extensive lakes and reservoirs. Breeding birds nest in trees, and are colonial – they will also join established nesting colonies of Grey Herons. They roost communally in trees or reedbeds in winter, sometimes with other heron species.

■ **Search tips** The Little Egret's size, bright white plumage and preference for open habitat makes it easy to spot. Scan over muddy harbours, wet grazing marsh and by lake islands. When resting in a hunched-up posture, it can look surprisingly small and could be mistaken for a gull at a casual glance. Birds sitting high in trees may give away the location of a nesting colony. In flight it may attract attention with a loud harsh call.

WATCHING TIPS

Little Egrets are interesting to watch and may be quite confiding. They can be quite active and vigorous when hunting, rushing after prey through shallow water. In the breeding season, they attract aggressive attention from nesting wetland birds, resulting in dramatic interactions. Numbers are high in late summer and autumn, and sizeable gatherings may form in suitable feeding grounds.

Super sites

1. Rutland Water and other Midlands reservoirs
2. RSPB Titchwell
3. RSPB Strumpshaw Fen
4. RSPB Minsmere
5. RSPB Ham Wall
6. RSPB Radipole Lake
7. Poole Harbour and Brownsea Island
8. RSPB Arne
9. Langstone Harbour
10. Pagham Harbour
11. RSPB Elmley Marshes
12. RSPB Dungeness

Grey Heron

This very familiar large wading bird can be found in low-lying areas around or near water almost everywhere in the British Isles, including at back garden ponds. It is more widespread in winter, as birds disperse from nest sites.

Length 90–98cm

| 1 | 2 | 3 | 4 | 5 |

J F M A M J J A S O N D

How to find

■ **Timing** Grey Herons are easy to see year-round, though they keep a slightly lower profile in early to mid-spring when incubating and caring for small chicks. Depending on their hunting success, long spells of the day are spent fairly inactive.

■ **Habitat** Most rivers and lakes will be visited daily by hunting Grey Herons. A good population of fish or other suitable prey and water shallow enough for wading is all that is required. They sometimes also hunt on open fields. Nesting colonies use stands of tall trees (not necessarily adjacent to water).

■ **Search tips** Because the Grey Heron's usual hunting style is to stand and wait for prey movement nearby, it does not often give itself away by movement. Scan along water edges for the hunched still shape of a heron searching for lunch. When not feeding, its chosen resting place may be well tucked into marginal vegetation, or somewhere as prominent as the crossbar of an electricity pylon. In spring and summer nesting wildfowl and waders may reveal a hiding heron by mobbing it out of cover.

WATCHING TIPS

Several city parks, including Regent's Park in London and Verulanium Park in St Albans, have breeding Grey Herons. Well used to the general public, these birds can be watched at very close range, though more rural heronries must be watched from a good distance to avoid disturbance. Activity at colonies is high in late winter as pairs court and build or renovate their nests. Despite spending much of their time doing very little, Grey Herons are fascinating to watch when actively hunting, and because of their liking for waterfowl chicks, often get involved in skirmishes with other wetland birds. If your garden pond has fish, a heron may well visit – discourage it by arching chicken wire over the water.

Purple Heron

Ardea purpurea

This shy heron is found over much of southern Europe and is a rare spring and summer visitor to Britain. Although declining over much of its range, it bred for the first time in Britain in 2010.

Length 78–90cm

1	2	3	4	**5**

J F M A M J J A S O N D

How to find

■ **Timing** The Purple Heron is a summer visitor, migrating to Africa for the winter. Most records in Britain are between April and August, with the later summer records often involving juveniles that have dispersed from mainland Europe.
■ **Habitat** This species needs shallow wetlands that have extensive reedbeds or other tall fringing vegetation around, both for feeding and for any breeding attempt (unlike the Grey Heron, it nests not in trees but in reedbeds). Dispersing young birds may visit less suitable-looking wetlands.
■ **Search tips** Because it is so shy and skulking, the Purple Heron is often well out of sight within the reeds when feeding, so most sightings are of flying birds. These present a similar general appearance to Grey Herons, so in spring and summer it is always worth giving any heron in flight a second look, particularly in suitable habitats in south and east England. The darker and more slender appearance of Purple Heron is apparent given a good view. When at a site where a Purple Heron is known to be present, carefully scan all reedy margins.

Super sites

1. RSPB Minsmere
2. RSPB Elmley Marshes
3. RSPB Dungeness

WATCHING TIPS

The breeding pair at RSPB Dungeness in 2010 was seen by many visitors, as the RSPB arranged a viewpoint from which they could be watched flying to and from the concealed nest at a safe distance. Visitors needed considerable patience – but there was of course a wealth of other wetland wildlife to enjoy at the reserve. Similar arrangements could be established if there are further breeding pairs at reserves in the future – and most sufficiently large reedy wetlands are on RSPB or other nature reserves. Otherwise, wandering juveniles are likely to be easier to watch than adults, as they will explore more varied habitat and may even feed in quite open areas.

Ciconia ciconia

White Stork

An unmistakable large bird, the White Stork is a rare but regular visitor to Britain from further south.

Length 100–115cm

| 1 | 2 | 3 | 4 | **5** |

J F M A M J J A S O N D

How to find it

■ **Timing** Sightings are most likely in mid-spring, when returning migrants 'overshoot' their breeding grounds during warm weather, and again in early to mid-autumn when dispersing young birds may wander to our shores.
■ **Habitat** White Storks feed on a wide range of prey, from small mammals and amphibians to insects and fish, and hunt on both open fields and around wetlands.
■ **Search tips** Finding a White Stork is mostly a matter of luck rather than judgement, but warm weather and southerly winds increase your chances. Birds will often pass overhead quite high, so scan the skies as well as suitable fields and wetlands. Some birds are found to be escapees from captivity.

WATCHING TIPS

Most British records of White Storks in spring are 'flyovers', which presumably correct their course and return south, perhaps without ever setting foot on British soil. Young birds in autumn are more likely to linger for a few days, and can then be watched hunting.

Plegadis falcinellus

Glossy Ibis

A unique and distinctive bird, the Glossy Ibis breeds in southern Europe but variable numbers visit Britain, mainly in autumn.

Length 55–65cm

| 1 | 2 | 3 | 4 | **5** |

J F M A M J J A S O N D

How to find it

■ **Timing** Autumn and early winter is the best time for finding Glossy Ibis in Britain, when birds are dispersing after the breeding season, though there may be arrivals in any month of the year.
■ **Habitat** Most birds visit marshland and wet pasture with flooded patches, probing wet ground for prey.
■ **Search tips** Suitable sites should be checked through autumn – scan the ground and bear in mind that birds may tuck themselves into hollows or by channels. Autumn arrivals often involve small flocks, so if you find one, check around for others.

WATCHING TIPS

Take care not to disturb feeding birds – they may have undertaken a long flight and be short of energy. Watch from a distance, or from a hide. Check any birds you see for leg rings, as there have been UK records of Spanish birds wearing colour rings.

Spoonbill

Platalea leucorodia

A spectacular large wading bird, the Spoonbill is a regular visitor to the UK, though in small numbers, and recently bred again in the UK (in both England and Scotland) after an absence of more than 300 years.

Length 80–93cm

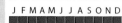

| 1 | 2 | 3 | **4** | 5 |

J F M A M J J A S O N D

How to find

■ **Timing** Although autumn and winter are probably best, you could see Spoonbills at any time of year. Flocks are more likely to visit after the breeding season, and may remain within the same general area for weeks, often 'touring' nearby sites on a regular daily routine.

■ **Habitat** Most Spoonbills in the UK stay at coastal sites, with marshland and shallow open water, either fresh or salt. However, they do also turn up further inland, usually at large lakes. They also seek out places with islands (within lakes, or sandbanks in estuaries) on which they can safely roost. Nesting is either in a tree or within a dense reedbed.

■ **Search tips** Spoonbills spend much time asleep, with their distinctive bills tucked away out of sight. Scan islands and sandbanks, and also large trees on islands, for a tall and sturdy white bird which may be sitting in full view. Also check for birds feeding in open water. They look very large in flight, and fly with the neck extended rather than retracted like an egret.

Super sites

1. RSPB Titchwell
2. Cley Marshes WT
3. RSPB Minsmere
4. RSPB Exe Estuary
5. Brownsea Island NT
6. RSPB Arne
7. RSPB Elmley Marshes
8. Oare Marshes

WATCHING TIPS

These birds are fascinating to watch, especially in flocks, with their unique feeding style and gregarious ways. You may need to wait patiently to see some action though as they spend hours of the daytime asleep. Flocks move in close synchrony, walking in lines as they feed, swishing their open bills through the water and snapping up small prey items. Then the flock takes off together and wheels around in a circle before heading off to the roosting ground. Spring birds sport golden breast-patches and may be seen flaring their crests in courtship display. Any sign of breeding behaviour should be reported to your local recorder, so that potential nest sites can be protected.

Pernis apivorus

Honey-buzzard

One of the most enigmatic birds of prey in Britain, the Honey-buzzard is both rare and rather shy. A few dozen pairs breed, mainly in the south and east but also in northern Scotland, and continental birds may pass through on migration.

Length 52–60cm
Wingspan 135–150cm

| 1 | 2 | 3 | 4 | 5 |

J F M A M J J A S O N D

How to find

■ **Timing** Honey-buzzards are late migrants. The easiest time to see them is when they have just arrived and are displaying on their territories at the start of the breeding season.

■ **Habitat** This is a forest bird, needing good tree cover in which to nest, but also plenty of glades and clearings where it can search for bee and wasp nests. Migrating birds are most often seen down the eastern coastline of Britain, where they will overfly all kinds of habitats – even towns and cities.

■ **Search tips** Because of the risk of disturbance, it is best to look for Honey-buzzards from established watchpoints, such as at Swanton Novers in Norfolk or Wykeham Forest in North Yorkshire. The second half of May is the best time, and the best weather for good views is fairly warm with a light breeze. Be aware that Common Buzzard is likely to be present at the same sites. In early autumn, keep an eye on the skies for migrating Honey-buzzards, but be aware that the risk of confusing a juvenile Honey-buzzard with Common Buzzard is even greater than with adults.

Super sites

1. Wykeham Forest
2. Clumber Park
3. Swanton Novers
4. New Forest

WATCHING TIPS

Witnessing the spring courtship display of the Honey-buzzard is high on the 'wish list' of many birdwatchers, but you will need a lot of luck. Making frequent visits to a watchpoint in late May is your best bet – visit early in the day if hot and still weather is forecast. The display includes dramatic climbs and dives, sometimes performed by the pair together. Should you find a potential Honey-buzzard in autumn, do your best to record as many details as possible, whether by taking photos or writing notes, to help get a definite identification. Keep an eye on local bird reports so you are ready to go out and find your own when there is a strong migration underway.

Red Kite

Milvus milvus

Once almost extinct here, the Red Kite is now returning to its former range across the whole of the UK following a series of reintroduction schemes. The areas around the reintroduction sites are still the best places to see it.

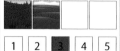

Length 60–66cm
Wingspan 175–195cm

1	2	3	4	5

J F M A M J J A S O N D

How to find

■ **Timing** Red Kites are present here all year round, though the most dramatic gatherings can be seen in winter. Some birds from the migratory populations further north-east may be seen arriving or departing along southern and eastern coastlines in spring and autumn.

■ **Habitat** Mature trees are required for nesting, and open fields provide ideal foraging habitat for food such as carrion and earthworms. Therefore, the ideal habitat for this species is farmland with scattered woodland, though in some areas it is increasingly visiting the outskirts of towns. As a carrion-feeder, it takes much road kill and is often seen flying by roadsides.

■ **Search tips** In core Red Kite country, finding the birds is usually simple, as they spend much time cruising around very visibly, scanning the ground below from high and low altitudes. The deeply forked tail eliminates confusion with other raptors, even from a distance. Productive feeding grounds may attract dozens of birds. They are unpopular with corvids, whose mobbing behaviour may give away the position of a perched kite. In general, though, Red Kites are not often seen perched, tending to settle in trees rather than more prominent spots.

Super sites

1. RSPB Ken-Dee Marshes
2. RSPB Carngafallt
3. Gigrin Farm
4. Rockingham Forest
5. RSPB Cwm Clydach
6. Aston Rowant
7. RSPB Otmoor
8. RSPB Church Wood

WATCHING TIPS

Watching this graceful bird on the wing is a real treat. It shows little concern for the presence of humans, and will glide by at close range and low altitude. Spending a few hours in one of the key areas should produce numerous encounters. Elsewhere, a sighting is always a possibility, especially in spring and autumn. There are a few 'Red Kite feeding stations' around the UK, most notably Gigrin Farm in Wales, where hundreds of kites (as well as Common Buzzards, Ravens and other scavengers) come to a daily handout of meat, giving truly spectacular views.

Haliaeetus albicilla

White-tailed Eagle

Our largest raptor owes its presence in Britain today to reintroductions, after it was exterminated in the early 20th century. Now, there are small populations in north-west and north-east Scotland and Ireland. Wandering young birds occasionally turn up elsewhere.

Length 70–90cm
Wingspan 200–240cm

| 1 | 2 | 3 | 4 | 5 |

J F M A M J J A S O N D

How to find

■ **Timing** This eagle may be seen at any time of year – this applies to immature birds as well as breeding adults. Like most large soaring raptors, it is usually not very active until a few hours after dawn, when there are warm air currents to help it gain height. Adults may be seen 'sky dancing' in spring.
■ **Habitat** The species' natural habitat is quite general, including low and high ground near the coast or inland lakes. In Scotland, the west coast population is based around quiet and rugged rocky coastlines and islands.
■ **Search tips** Relatively warm and still days are most conducive to White-tailed Eagle aerial activity, when scanning the skies in suitable habitat should eventually result in sightings. You may also see birds making shorter flights low over the cliffs. Golden Eagles occur in the same habitat, but have a more elegant outline compared to the short-tailed, 'plank-winged' White-tailed. Resting birds often pick a visible perch, their hulking outline breaking the skyline on a high rock, or looking oversized in a dead tree. Whale-watching boat trips out of west coast fishing villages can be good for seeing White-tailed Eagles. If a wandering youngster is found near your area, it may well hang around for weeks – search on fine days.

Super sites

1. Gairloch Harbour
2. RSPB Glenborrodale
3. RSPB Mull Eagle Watch

WATCHING TIPS

The White-tailed Eagle population in Scotland is still small and fragile, so it is vitally important that breeding pairs are not disturbed. On Mull, there is a dedicated eagle-watching hide overlooking Loch Frisa, from where you can safely watch a nesting pair of White-tailed Eagles and their chicks between August and early September – check the RSPB's 'Date with Nature' pages for details. Some of Scotland's White-tailed Eagles are fitted with colour-coded wing tags to help track their movements. You can report sightings of tagged birds through the Euring website.

Marsh Harrier

Circus aeruginosus

On the brink of extinction in the UK in the early 1970s, the Marsh Harrier has staged a remarkable recovery, thanks to protection and regeneration of its marshland habitat. It is now spreading from its East Anglian stronghold.

Length 48–56cm
Wingspan 115–130cm

1	2	3	4	5

J F M A M J J A S O N D

How to find

■ **Timing** Marsh Harriers are partially migratory in the UK, with those breeding further north heading south in autumn. However, there are birds present in the southerly parts of the range, in particular Kent, throughout the year.

■ **Habitat** These harriers are closely tied to wetland habitats, with extensive reedbeds and marshland with shallow pools (with breeding waterfowl) offering ideal hunting and nesting conditions. They will also hunt over rough wet pasture and along rivers. When on migration they may be seen overflying any habitat type, usually at much higher altitude than the usual hunting flight.

■ **Search tips** In suitable habitat, Marsh Harriers are usually easy to see. Their hunting flight or 'quartering' is close to the ground, often barely skimming the tops of the reedbeds. Find an elevated spot so you can scan a wide area of marsh (many good sites are nature reserves with ideally placed hides) and look for a long-tailed, slow-flying large bird of prey. Migrating Marsh Harriers are easily overlooked as they go over very high, and in soaring flight could be confused with Common Buzzards and other raptors.

Super sites

1. RSPB Loch of Strathbeg
2. RSPB Leighton Moss
3. RSPB Titchwell
4. RSPB Minsmere
5. RSPB Elmley Marshes
6. RSPB Dungeness

WATCHING TIPS

A visit to a marshland RSPB reserve in the south-east should provide numerous Marsh Harrier sightings, and the opportunity to watch hunting birds at length as they slowly patrol their habitat, before hovering and dropping quickly down into the reeds. In early spring, the courtship display may be seen, with both members of a pair performing roller coaster aerobatics. Courtship also includes mid-air food passes from male to female, which continues through the breeding season. In winter, communal roosts form in reeds or bushes, with birds arriving from mid-afternoon.

Hen Harrier

This lovely bird's population in the UK is under constant pressure from relentless illegal persecution. It breeds in the remote uplands of Scotland, Ireland and Wales, but is barely hanging on in England. It moves to lowlands and coasts for winter.

Length 44–52cm
Wingspan 100–120cm

| 1 | 2 | 3 | 4 | 5 |

J F M A M J J A S O N D

How to find

■ **Timing** Hen Harriers return to their breeding grounds from late March, and most or all will have left by late September. In the winter months they may wander quite widely, with highest numbers on the south and east coasts.
■ **Habitat** The usual breeding habitat is upland heather moor, though in other parts of its range it uses crop fields, rough grassland and young conifer plantations, and this habit may become widespread in the UK in the future. In winter, it favours low-lying coastal marshland and rough pasture, but may also hunt over heaths and arable farmland.
■ **Search tips** Like other harriers, it is a low-level hunter, so scan at ground level. The female's large white rump patch is noticeable from a distance, while the male's pale grey plumage with black wing-tips can give the impression of a gull at first glance. Also check along lines of fence posts as Hen Harriers sometimes employ the 'sit and wait' hunting method, especially on windy or very still days, which don't suit the usual breeze-assisted quartering flight.

Super sites

1. RSPB Insh Marshes
2. RSPB Hoy
3. RSPB Coll
4. RSPB Trumland
5. RSPB Geltsdale
6. RSPB Strumpshaw Fen
7. RSPB Minsmere
8. RSPB Elmley Marshes

WATCHING TIPS

Hen Harriers 'sky dance' in spring, a dramatic display of aerobatic skill which can be watched at various RSPB reserves in Scotland. As the breeding season progresses, the male supplies the female with food via mid-air food passes. Any sign of breeding activity in England away from the Forest of Bowland in Lancashire, the only regular England breeding site, should be reported to your local bird recorder to help ensure the birds are protected. Hen Harriers in winter sometimes go to communal roosts at dusk, and may join Marsh Harrier roosts.

Montagu's Harrier

Circus pygargus

This elegant harrier is particularly difficult to see. It is a summer visitor, and pairs don't return to the same territory year on year, so it is unpredictable as well as very rare. Migrants travel along the south or east coasts.

Length 43–47cm
Wingspan 105–120cm

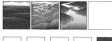

1	2	3	4	5

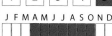

J F M A M J J A S O N D

How to find

■ **Timing** It returns to its breeding grounds in April or May, and departs again by October.

■ **Habitat** Like other harriers, this is a bird primarily of open landscapes. It nests on the ground, so requires quite thick vegetation at ground level in which to conceal itself and its nest. In the UK, most Montagu's Harriers nest in crop fields. On migration, it may stop off at low-lying coastal sites with marshland or rough pastures.

■ **Search tips** If your usual birdwatching route includes arable farmland, there is always a small chance of finding nesting Montagu's Harriers. They hunt in typical harrier style, with a slow, low, wafting flight, gaining lift by flying into the wind. On days when the weather is not conducive to this they will watch for prey from a perch a metre or two above the ground. Scan over fields and along rows of posts, and check gates and other suitable-looking low perches. Migrating birds may go over at some height. Beware confusion with Hen Harriers in spring and autumn – the two species may even occur together.

Super sites

■ Any reserves on the east coast of England may produce sightings in spring or autumn

WATCHING TIPS

This bird's breeding population is so small and vulnerable that it is very rare for nest sites to be publicised, but it is also important that the landowners are made aware of the birds' presence to avoid accidental disturbance. If you find Montagu's Harriers in the breeding season, tell your local bird recorder, who will ensure the right people are informed. Occasionally, Montagu's Harriers will nest somewhere where safe public access for visitors can be arranged – check the RSPB's Date with Nature pages through the summer. Look out for migrants along the east and south coasts in autumn.

Goshawk

This spectacular but secretive large raptor is on the increase in the UK. At present it has a wide but patchy distribution in England, Scotland and Wales, with most of the population living in large tracts of undisturbed woodland.

Length 48–62cm
Wingspan 135–165cm

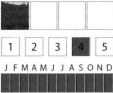

1	2	3	4	5

J F M A M J J A S O N D

How to find

■ **Timing** While Goshawks are present (with adults staying in the same general area) year round, the best time to look for them is in late winter to early spring, when they are performing their display flights.

■ **Habitat** In the UK, this is a forest bird, occupying both deciduous and coniferous woodland in lowlands and uplands, and favouring areas with extensive continuous tree cover. On the continent, it is increasingly becoming a bird of the suburbs, breeding in town parks, and this could occur in the UK in due course as the population increases.

■ **Search tips** To find displaying Goshawks, pick clear and reasonably still days between late February and early April, and find a viewpoint that gives you a wide area of sky to scan. A telescope will be helpful. The four hours either side of midday sees the most activity. Be aware that there will probably be other soaring birds of prey in the air, especially Sparrowhawks and Common Buzzards. At other times of year, encountering a Goshawk is very much a matter of luck, though fine weather could induce them to soar and circle over their territories at other times of year.

Super sites

1. Ladybower Reservoir
2. RSPB Lake Vyrnwy
3. RSPB Nagshead
4. Forest of Dean
5. RSPB Garston Wood
6. New Forest

WATCHING TIPS

The spring display can be quite exhilarating to watch, with a spiralling 'sky dance' performed by a pair together, and various other aerobatics. More usual, however, is simple high circling over the woodland, which serves as an advertisement of their presence to other Goshawks. The species is very vulnerable to persecution, so any sightings away from known strongholds should be reported. Many reported Goshawks, however, are in fact misidentified Sparrowhawks, so take care to make detailed notes or take photographs.

Sparrowhawk

Accipiter nisus

Its numbers declined sharply during the DDT era in the second half of the 20th century, but the Sparrowhawk has bounced back and is now a common and widespread raptor, and the likeliest to be seen in a garden.

Length 28–38cm
Wingspan 55–70cm

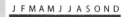

| 1 | 2 | 3 | 4 | 5 |

J F M A M J J A S O N D

How to find

■ **Timing** Sparrowhawk activity can be seen throughout the year. On fine days in late winter and early spring, territorial adults are often easy to see as they circle high over their territories.

■ **Habitat** This raptor is well adapted to chase its prey among dense tree cover, so its main habitat is woodland (especially deciduous), but it is also at home in parkland, and also among gardens provided there are trees nearby – large mature trees are required for nesting.

■ **Search tips** If you have a garden, it is likely to be visited by your local Sparrowhawk from time to time, especially if you have good numbers of smaller garden birds visiting. This may be a flying visit (if birds at a feeding station suddenly scatter, this is a good sign that a Sparrowhawk is nearby); also look out for birds waiting in well-hidden spots among the foliage. On warm days soaring Sparrowhawks may be seen over almost any wooded habitat, and hunting birds sometimes fly low along hedge-lined country lanes ahead of cars, ready to catch any small bird flushed by the vehicle.

WATCHING TIPS

Not everyone is pleased to see Sparrowhawks visiting their garden, but the presence of these predators indicates a healthy and productive local bird population, and there is no need to fear that Sparrowhawks will wipe out the smaller garden birds. Many householders have witnessed a Sparrowhawk making a kill in their garden – a dramatic spectacle. In spring, pairs of Sparrowhawks display above their territories, and you may witness chases and other aerobatics. Female Sparrowhawks are much larger than males and often mistaken for Goshawks – the more familiar you are with the commoner species, the more likely you'll be to recognise a Goshawk when you see one.

Common Buzzard

Once only common in Scotland, Wales and the west of England, this species has spread and increased since the late 20th century. Today it is probably our commonest raptor, and can be seen almost everywhere, though remains rare in western Ireland.

Length 51–57cm
Wingspan 113–128cm

1	2	3	4	5

J F M A M J J A S O N D

How to find

■ **Timing** Common Buzzards are present year-round in Britain, with adults remaining on or near their breeding grounds all year. They need thermals to soar, so are not usually seen flying high until mid-morning.

■ **Habitat** Most open countryside holds Common Buzzards, whether it is lowland fields, scattered woodland or high heather moor. Provided there are at least a few mature trees or inaccessible crags for nesting, Common Buzzards are likely to be present. Roaming young birds may also be seen away from even these habitats, for example over urban areas and flat, bleak coastlines.

■ **Search tips** On reasonably clear, still days, Common Buzzards will take to the skies four or five hours after dawn to soar and search the ground below for prey. Scan for a broad-winged, fan-tailed shape wheeling in slow circles. When the weather is not suitable for soaring, the birds are often seen on low perches or on the ground, sitting still and watching for prey. They may join flocks of gulls following the plough. When flying purposefully they flap strongly and hold their tails closed. This makes the tail look longer – beware confusion with other species, especially Honey-buzzard.

WATCHING TIPS

In the right time and place you can watch multiple Common Buzzards riding the thermals for long spells, though views may be distant. When perched on or near the ground they are wary, so using a hide (including a car) may be helpful. In many parts of their range they will share their airspace with other raptors – Red Kites, Marsh Harriers, even Golden Eagles. Therefore familiarising yourself with the shape, behaviour and wide range of plumage variation of Common Buzzards is essential as it will help you to pick out more unusual large birds of prey.

WATCHING BIRDS OF PREY

Birds of prey hold a special place in most birdwatchers' hearts. With their skill, power and spirit they seem to possess all the qualities that draw us to birds in general. Unfortunately, they have long been persecuted just for being predators, and it is only since the mid-20th century that they have enjoyed proper legal protection. Nevertheless, some illegal persecution still continues. Watching birds of prey requires an understanding of their way of life, and an appreciation of the threats that they still face.

On the air

Weather conditions have a great deal to do with bird of prey activity. Many species search for prey from a high soaring flight, but reaching these heights is energetically expensive if done through flapping flight. Therefore, larger raptors like buzzards and eagles use thermals (rising air currents) to gain height. Significant thermals don't begin to develop until the sun has been up for several hours, so you are unlikely to see soaring large raptors until mid-morning or midday (depending on the time of year), and on miserable rainy days they may not fly at all.

Wind speed also affects bird of prey activity. In this case, it is the species that hunt from a low-level flight that are more affected – this includes the harriers and also open-country owls like Barn and Short-eared. These birds fly into the wind, using it for uplift. They can thus hover and 'hang' on the breeze more easily. Hovering Kestrels also use the wind in a similar way. If there is next to no wind, these species are more likely to hunt from a perch.

Dead giveaway

Every so often, you'll be watching a flock of birds feeding – perhaps waders on an estuary, or finches around a bird table. All of a sudden, the flock takes flight. The waders swirl up into the sky, the finches scatter into nearby bushes. The reactions carry the same message – a bird of prey has arrived on the scene, and if you quickly search the skies you should be able to find it.

The behaviour of other birds often reveals the presence of a raptor or owl, although different species elicit different reactions. A flock of feeding waders will ignore a Kestrel hovering nearby, but will take to the air in a panic if a Buzzard flies over. The reaction isn't always one of fear, either. Most birds of prey will be mobbed by other birds from time to time, with crows, gulls and hirundines especially determined mobbers of raptors in flight, while a perched Tawny Owl will provoke mobbing by a whole range of small woodland birds. Wherever you are birding, keep alert for these reactions as they can bring your attention to a bird of prey. If you are very lucky, you could even witness a strike from the predator.

Kestrel

Golden Eagle

Barn Owl

SIGNS OF TROUBLE

It is a great shame, in all senses of the word, that the illegal killing of birds of prey is still such a major problem in Britain. It is because of persecution that England's population of Hen Harriers is on the brink of extinction, and Golden Eagles in Scotland are still absent or very scarce in large areas of suitable habitat. Methods of killing include shooting, poisoning and trapping, and destruction of nests.

If you should come across a dead bird of prey, a snare or trap, or what you think may be poisoned bait, contact the local police and take photographs if possible. Some kinds of bird traps are legal (for example Larsen cage traps when used to trap legal 'pests' like some species of crows), but strict guidelines cover their use, so don't be afraid to seek advice.

Should you find the nest-site of an uncommon bird of prey, don't be tempted to tell other birdwatchers or publicise your sighting. These birds are too vulnerable to take any risks – not only are they at risk of persecution, but they are also often targeted by egg collectors. Inform your local recorder, but otherwise keep the details private.

Hen Harrier

Rough-legged Buzzard

Buteo lagopus

The Rough-legged Buzzard breeds in and near the Arctic Circle, migrating south in winter. A few visit Britain each winter, mainly along the east coast – sometimes numbers barely reach double figures but in good years more than 100 arrive.

Length 50–60cm
Wingspan 120–150cm

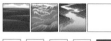

| 1 | 2 | 3 | 4 | 5 |

J F M A M J J A S O N D

How to find

■ **Timing** Mid-winter is the best time to look. Keep an eye on weather forecasts for Central and Eastern Europe, the main wintering range, as severe conditions here will push more birds westwards. Like Common Buzzards, they use thermals to soar so most activity is around the middle of the day.

■ **Habitat** All kinds of open countryside may be used for foraging. Many UK sightings are right on the coast, on rough grassland or marshland, but it may also venture inland to upland moors.

■ **Search tips** Finding Rough-legged Buzzards requires similar strategies to finding Common Buzzards. Although arrivals are unpredictable, certain areas do have a good track record for attracting the birds, and Rough-legged Buzzards tend to remain within a relatively small area once they have arrived. Choose a likely area and a fine day, and be prepared to put in some hours at a suitable watchpoint which allows you to scan a wide vista of land and sky. Any sightings should be carefully documented, with photographs if possible, to make certain of identification. A good view of the legs is particularly helpful, as these are feathered (bare in Common Buzzard).

Super sites

1. Wells-next-the-sea
2. Horsey area
3. RSPB Elmley Marshes
4. Capel Fleet

WATCHING TIPS

With good and prolonged views, you can begin to discern the subtle differences between this species and its much more common relative, the Common Buzzard. The two species can usually be separated by plumage differences, but they also show different behaviour – for example, the Rough-legged habitually hangs and hovers in a headwind when hunting, and spends less of its time soaring high. If you see the two species together you may also note the Rough-legged's slightly larger size and proportionately longer wings with a more relaxed-looking flight action.

Golden Eagle

Aquila chrysaetos

This majestic but much persecuted bird is found in the rugged Highlands of Scotland, especially in the west. At the time of writing, a solitary male still holds territory in the Lake District, and a reintroduction project is under way in Ireland.

Length 75–88cm
Wingspan 204–220cm

1	2	3	4	5

J F M A M J J A S O N D

How to find

■ **Timing** Adult Golden Eagles stay around their breeding territories throughout the year. Youngsters wander widely in their first year or two before settling down on a territory of their own. They need thermals to soar, so are most likely to be seen on the wing in the warmest hours of the finest days.

■ **Habitat** High open moorland, the lower slopes of mountains, and high steep pasture is typical Golden Eagle habitat – the less disturbed by human activity, the better. Some of the highest numbers are found on the islands off the north-west coast of Scotland. A breeding territory needs to have the odd tree or crag for a nesting place – nests are often used for many years in succession.

■ **Search tips** Whenever you are in Golden Eagle country, keep an eye on the skies and scan along the ridges and crests of mountains for soaring birds. Beware confusion with Common Buzzards, which are likely to be present in much higher numbers in all Golden Eagle habitat. Finding a perched eagle is a tall order, and only likely to happen if you are walking in the mountains – look for a very bulky shape perched among rocks or in a tree.

Super sites

1. Findhorn Valley
2. RSPB Glenborrodale
3. Mull
4. RSPB The Oa
5. RSPB Haweswater

WATCHING TIPS

Views of Golden Eagles are often frustratingly fleeting – perhaps just a glimpse of a long-winged shape appearing and disappearing as it glides low along a mountain ridge. Typical Highland weather conditions may also put paid to your ambitions, so if seeing a Golden Eagle is your goal, you may need to allow several days in the right areas. At the time of writing, the lone male Golden Eagle at RSPB Haweswater in the Lake District displays for weeks each spring, trying to attract a new mate, and visitors in spring have a good chance of watching his dramatic rolling courtship flight from the RSPB viewpoint. He may be seen at other times of year, but is much less reliable when not displaying.

Osprey

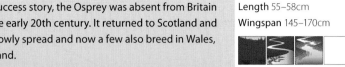

Pandion haliaetus

Another raptor success story, the Osprey was absent from Britain for decades in the early 20th century. It returned to Scotland and from there has slowly spread and now a few also breed in Wales, Ireland and England.

Length 55–58cm
Wingspan 145–170cm

| 1 | 2 | 3 | **4** | 5 |

J F M A M J J A S O N D

How to find

■ **Timing** Adult Ospreys start to return to their nests in mid-March. On southbound migration, they make leisurely progress, beginning in late August, and in early autumn many (including birds from the near continent) make long stopovers at water bodies all over Britain. Daily activity is not highly dependent on air temperature.

■ **Habitat** Ospreys nest in tall trees, which may be isolated or within dense stands of woodland. They will also use artificial breeding platforms. Nests may or may not be alongside water, but there needs to be fish-rich rivers or lakes fairly near.

■ **Search tips** Many Osprey nests are on protected sites and have viewing facilities for visitors. Seeing the birds is therefore very simple, although you may need to wait a while before you see any interesting activity, especially in the middle of the season when the adults are incubating or brooding small chicks. From late summer to mid-autumn, any lake may have a visiting Osprey, and flyover birds may be seen along the east coast especially as they track south. Returning migrants in spring tend to make much quicker progress with few or no stopovers.

Super sites

1. RSPB Loch Ruthven
2. Rothiemurchus Trout Fishery
3. RSPB Loch Garten
4. Loch of the Lowes
5. RSPB Wales Osprey Project at Glaslyn
6. Rutland Water

WATCHING TIPS

A day at a reserve with nesting Ospreys, such as the RSPB's Loch Garten, should result in some good sightings. Visiting early in the spring will give you views of the pair establishing their bond and perhaps dealing with challengers, while a visit in mid- to late summer will hopefully give you the chance to watch well-grown chicks competing for fish deliveries and exercising their growing wings. Some fish farms offer visitors the chance to watch and photograph hunting Ospreys for a fee (to offset the loss of trout to the birds!).

Falco tinnunculus

Kestrel

Once our commonest bird of prey, the Kestrel is currently declining rather rapidly for reasons that are unclear, but it remains very widespread. It can be seen anywhere in Britain, and its unique habits make it highly distinctive.

Length 32–35cm
Wingspan 71–80cm

| 1 | 2 | 3 | 4 | 5 |

J F M A M J J A S O N D

How to find

■ **Timing** Kestrels can be seen all year round and may be active at any time of day. Recently fledged young birds in early summer often give close views.

■ **Habitat** This raptor catches most of its prey on the ground, so needs open country for hunting. This may be pasture, heathland, woodland glades, parkland or similar habitats – anywhere with a good population of small mammals. Kestrels frequently hunt along grassy motorway verges, river banks and the set-aside edges of arable fields. They need high ledges or hollows for nesting, whether in trees, on cliff edges or on buildings.

■ **Search tips** A hovering Kestrel immediately draws the eye, flapping hard to hold position several metres above the ground. Kestrels also often hunt from perches, and have a distinctive upright but hunched posture as they sit on a post, telegraph pole or other elevated spot. Whenever in open countryside, check the sky and any obvious looking perching spots. Town Kestrels hunt more birds than mammals and may visit gardens, but a bird of prey in a garden is much more likely to be a Sparrowhawk.

WATCHING TIPS

The full process of a Kestrel hunt can often be watched at your leisure. The bird begins with a high hover, then drops a few metres to a lower position and hovers again, homing in on its prey. The final hover may be just a couple of metres above ground level, before the bird drops into the grass. No other British raptor has this hunting style. In normal flight and when soaring, the Kestrel may be confused with other raptors, but is has a longer tail than other falcons and has longer, narrower wings than the Sparrowhawk. It will also usually give itself away eventually by hovering. Newly fledged kestrels will sit in plain view, waiting for a prey delivery from a parent.

Merlin

Falco columbarius

This highly dynamic falcon is our smallest raptor. It is quite widespread throughout Britain but scarce and may be declining. It nests on uplands but moves to lower ground, usually on or near the coast, for winter.

Length 25–30cm
Wingspan 50–62cm

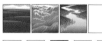

| 1 | 2 | **3** | 4 | 5 |

J F M A M J J A S O N D

How to find

■ **Timing** Merlins are on their moorland breeding grounds between mid-April and early October. Mid-winter is the best time to seek them out along the south and east coasts.

■ **Habitat** The typical breeding habitat is high, undisturbed, undulating heather moorland with a good population of small birds, especially Meadow Pipits and Skylarks. It may also nest in thick, rough grassland or young conifer plantations. The nest is usually on the ground among dense vegetation, so trees and crags are not necessary. Wintering grounds are more low-lying and often coastal, with rough pasture, maritime heath and marshland all used.

■ **Search tips** Hunting Merlins are often seen only by chance and briefly, as they chase a small bird at very high speed close to the ground. At all times in flight they move with great speed and purpose, and are difficult to follow. When in Merlin habitat keep looking all around you at close to ground level. Scan all potential perches – rocks, hummocks, fence posts and so on. They may be seen near dusk flying into communal roosts in reedbeds or other dense cover.

Super sites

1. RSPB Trumland
2. RSPB Forsinard Flows
3. RSPB Marshside
4. Martin Mere WWT
5. RSPB Blacktoft Sands
6. RSPB Northward Hill
7. RSPB Elmley Marshes
8. RSPB Pulborough Brooks

WATCHING TIPS

Although hunting Merlins are not easy to watch, sometimes the prey will lead the chase up into higher air and then amazing views can be enjoyed and the Merlin's fast and agile pursuit can be fully appreciated – as well as the often impressive escape attempts of its prey. Skylarks in particular attempt to draw a pursuing Merlin into a more aerial chase. Merlins sometimes disguise their predatory intent by adopting a radically different flight style – a leisurely, bounding flight reminiscent of a Mistle Thrush. This may escape the attention of potential prey until the Merlin gets close enough for a final chase. Merlins will also sometimes chase and mob larger raptors.

Hobby

One of the few raptors that visit us in summer only, the elegant and graceful Hobby has increased its range in the 21st century. It now breeds patchily across most of England, and edges into east Wales and south-east Scotland.

Length 30–36cm
Wingspan 82–92cm

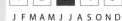

1	2	3	4	5

J F M A M J J A S O N D

How to find

■ **Timing** Hobbies arrive on the south and east coasts in mid-spring and often form large gatherings at sites close to the coast before dispersing to their breeding grounds. They depart in early or mid-autumn. Warm days with more insect activity encourage more Hobby activity.

■ **Habitat** This is a bird of fairly open habitat, though it needs trees for nesting (it uses the old nests of other birds such as crows). Lowland heath, farmland with scattered copses and pasture around lakes and reservoirs are all good places to look. Newly arrived migrants often gather around coastal marshes – the same sort of habitat that attracts migrating Swallows and martins.

■ **Search tips** Hobbies catch their prey in the air, often well above ground. Check the skies when in suitable habitat for a scythe-winged bird cruising and hawking gracefully and at rather low speed in wide circuits. Dragonflies are favourite prey and any pool or stream with dragonflies around may draw in a Hobby. In more purposeful directional flight it moves very fast and could be mistaken for a Kestrel or other falcon, flying with fast-flickering wings.

Super sites

1. RSPB Middleton Lakes
2. RSPB Otmoor
3. RSPB The Lodge
4. RSPB Lakenheath Fen
5. RSPB Arne
6. New Forest
7. Thursley Common
8. RSPB Rye Meads
9. Stodmarsh NNR
10. RSPB Dungeness

WATCHING TIPS

Watching a Hobby hunting is impressive. It catches and eats insect prey on the wing, without breaking stride – the prey is snatched in the talons which are then brought up to the bill. You may even see the insect's wings spiral downwards as the Hobby removes them. It is also an expert hunter of birds, even species as aerially skilled as Swallows and Swifts. It rarely engages in sit-and-wait hunting, but may be seen perched in a tree or on a post in between hunting forays. If you encounter a gathering of Hobbies in spring, check carefully as you could be lucky enough to find a rare Red-footed Falcon associating with them.

Peregrine Falcon

Falco peregrinus

With a very chequered history in Britain, the Peregrine is one of our best-known birds and is currently doing well here, though persecution is a constant worry. It is found around most of the coastline and patchily inland.

Length 36–48cm
Wingspan 95–110cm

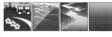

1	2	3	4	5

J F M A M J J A S O N D

How to find

■ **Timing** Peregrines can be seen year-round. Some upland birds may move to lower ground in winter, but most adult pairs remain near their nest site all year, while young birds wander widely for a year or two. Early spring is perhaps the best time to see them as territory-holders are especially visible and vocal.

■ **Habitat** Adult Peregrines base themselves around a nest site, usually on a cliff, crag or building, which may be used by successive pairs for decades, but their hunting territory extends for miles around. Their hunting style suits quite open countryside – coastal marshes, cliffs with nesting seabirds, farmland, moor – anywhere there are good concentrations of medium-sized birds.

■ **Search tips** Early in the breeding season, territorial birds give themselves away with their loud shrieking calls and aerial patrols – the distinctive anchor shape in flight is unlike any other largish raptor. Recently fledged young birds are also very noisy. Along coastlines with steep cliffs, scan the cliff face when possible for birds flying past or resting on a ledge. Any mass panic among a flock of ducks or waders could indicate the presence of a hunting Peregrine – search the sky above the flock. In open countryside, check pylons for resting Peregrines.

WATCHING TIPS

Watching Peregrines from a clifftop can be very rewarding, as you should get top-down or even eye-level views. Be extremely careful on unfenced cliff edges, though! Most urban nest sites are well known and some have viewpoints set up from which the birds can be watched at and around the nest. In winter, marshlands that attract flocks of ducks and waders will also attract Peregrines and you could witness the spectacular 'stooping' attack. The kill is usually dealt with on the ground (though in the breeding season most kills will be carried back to the nest).

Rallus aquaticus

Water Rail

This shy bird of well-vegetated wetlands is quite widespread in Britain, especially the south, but can be a real challenge to see. In winter, our breeding population is supplemented by arrivals from the near continent, in numbers depending on the weather.

Length 23–28cm

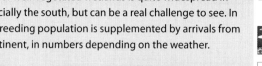

| 1 | 2 | 3 | 4 | 5 |

J F M A M J J A S O N D

How to find

■ **Timing** Although Water Rails are present all year round, they are usually considerably easier to see in winter. This is partly because of increased numbers, but also because freezing water limits the places they can forage and may force them into the open.

■ **Habitat** This species is most often found in fresh or brackish wetlands with shallow water and plenty of thick emergent vegetation. It nests within this thick vegetation – in winter it may make do with smaller pools and may forage a short distance away from water. Immigrants from the continent sometimes turn up in unexpected places, including gardens.

■ **Search tips** Watching from a well-placed hide is probably the best way to see this bird, as (in most areas) it is shy, stays close to cover and is quick to disappear at the slightest disturbance. Keep checking along the edges of vegetation and along channels between stands of reeds. The distinctive pig-like squealing call indicates the bird's presence, though does not guarantee a sighting. In some places Water Rails come to feeding stations in winter.

WATCHING TIPS

The usual view of a Water Rail is a hunched dark bird creeping furtively along the edge where a bank of reeds or sedges meets the water, and hurrying out of sight the moment it realises you are there. Watch patiently and quietly from a hide and you may see it pursuing insects, or swimming across short stretches of open water. If watching a feeding station near water in winter, check the ground from time to time for Water Rails picking up fallen scraps. In freezing weather, Water Rails are likely to be much more visible as long as there are some patches of unfrozen water and wet ground still available.

Spotted Crake

Porzana porzana

This is one of our rarest breeding birds, and one of the shyest, which makes it difficult to assess just how many visit Britain in the summer and how successfully they breed. It has a scattered distribution in Scotland and England.

Length 22–24cm

1	2	3	4	5

J F M A M J J A S O N D

How to find

■ **Timing** Spotted Crakes are on territory by mid-April and depart in September. For the best chances of an encounter, though not necessarily a sighting, be at a suitable site for dusk, when the males begin to give their territorial calls.

■ **Habitat** Like the Water Rail, this species needs well-vegetated wetlands. The more extensive the area of habitat, the more chance of it supporting breeding birds, though dispersing youngsters and southbound migrants may make do with smaller areas.

■ **Search tips** The male's whip-crack call, given at night from a well-concealed spot among the vegetation, is the only obvious sign of the presence of a Spotted Crake. Migrants do not call, and so their discovery is the result of patience, luck and choice of a good viewpoint at a suitable site. Although they may turn up in almost any patch of good habitat, there are certain favoured sites where several birds visit most years. You are also more likely to be lucky if you search at dawn and/or dusk as this is when the species is most active.

Super sites

1. RSPB Insh Marshes
2. Porthellick Pool on St Mary's, Scilly
3. RSPB Marazion Marsh
4. Somerset Levels
5. RSPB Ouse Washes

WATCHING TIPS

As with Water Rails, you should ideally use a hide or other well-concealed spot when watching for Spotted Crakes, as they are shy and easily spooked. When they are not disturbed they will move between vegetation and open water, stepping along methodically most of the time but with occasional short dashes after prey, and they also swim competently. As the Spotted Crake could be confused with other, much rarer vagrant species, in particular the Sora, a North American species, it is always a good idea to take field notes and/or photographs if possible especially if you are away from regular Spotted Crake sites.

Corncrake

Once a common farmland bird throughout Britain, the Corncrake is now restricted to north-west Scotland and Ireland, but its fortunes are now improving thanks to more sympathetic farming methods. There is also a reintroduction project under way in eastern England.

Length 27–30cm

| 1 | 2 | 3 | **4** | 5 |

J F M A M J J A S O N D

How to find

■ **Timing** The Corncrake is a summer visitor, on its breeding grounds between mid-spring and early autumn. Migrants may turn up away from the usual sites, but are seldom seen because of their very shy habits.

■ **Habitat** This bird nests in long hay meadows or similarly dense crop fields, on low-lying ground. It only thrives where the timing and method of harvesting follows Corncrake-friendly methods, so that the young birds are fully mobile before the field is cut, and cutting from the inside outwards to give them an escape route. Unlike the other rails and crakes it has no affinity with water.

■ **Search tips** Male Corncrakes give their rasping two- or three-note call through the night and sporadically in the day, a clear indication of the presence of breeding birds. Though they spend much time feeding out of sight in long vegetation, they are not overly shy and may call from visible places – try driving slowly around the lanes in suitable habitat, scanning the field edges as you go. Sometimes just the calling male's uptilted head is visible over the grass.

Super sites

1. RSPB Coll
2. RSPB Balranald
3. Inner Hebrides

WATCHING TIPS

Watching Corncrakes presents the same sorts of challenges as watching other crakes and rails – they are usually hidden in vegetation. Picking a somewhat elevated viewpoint over fields with some areas of bare or more open ground improves your chances of longer views, but avoid entering fields in an attempt to flush out a calling bird. On some of the Inner Hebrides the populations are quite dense and the birds rather unafraid of people, so it may be just a matter of patience before good and prolonged views are had. Migrant Corncrakes may give good views if they happen to turn up in spots with little ground vegetation.

Moorhen

Gallinula chloropus

One of the typical birds of park lakes, the Moorhen is a common, confiding and easily observed water bird.

Length 32–35cm

| 1 | 2 | 3 | 4 | 5 |

J F M A M J J A S O N D

How to find it

■ **Timing** Moorhens can be watched year-round. They have chicks in tow from mid-spring.

■ **Habitat** Any kind of low-lying, well-vegetated fresh water will have breeding Moorhens, even quite small ponds. They also forage in nearby fields and may visit gardens close to water, especially in winter.

■ **Search tips** It is usually easy to find Moorhens. They tend to swim quite close to the shore or islands, spend much time feeding among grass near the water's edge, and may perch in trees. They give frequent loud, liquid calls. Nests are anchored to vegetation and quite well hidden.

WATCHING TIPS

Moorhens become very tame in town parks and can be watched at close range. Both adults and older siblings care for the small chicks, which follow their parents around with squeaky begging calls. They will visit feeding stations to take scraps from the ground.

Coot

Fulica atra

A very common water bird, the Coot occurs throughout Britain except the far north-west, on all kinds of low-lying water.

Length 36–38cm

| 1 | 2 | 3 | 4 | 5 |

J F M A M J J A S O N D

How to find it

■ **Timing** This bird can be seen at any time of year. Large gatherings form on more substantial lakes in winter.

■ **Habitat** It is found on most medium to large water bodies, including deeper and less well-vegetated waters. It is one of the first species to colonise newly flooded gravel workings.

■ **Search tips** Coots are often found out in the middle of the lake where they dive to pull up underwater vegetation, and are gregarious (frequent fights notwithstanding) all year round. In town parks they are very approachable and you can watch their breeding behaviour easily.

WATCHING TIPS

Town Coots can easily be watched at close range all year, and you can observe their breeding cycle, including the unusual aggression sometimes directed by parents towards small chicks. The adults themselves are often aggressive to each other and other waterfowl.

Grus grus

Crane

This giant among birds is an extremely rare breeding species in Britain. After an absence of some 400 years, a few pairs returned in the late 20th century to nest at a single site in East Anglia. Plans are under way for reintroductions.

Length 110–120cm

1	2	3	4	5

J F M A M J J A S O N D

How to find

■ **Timing** There are some 30 Cranes in the East Anglian population. They are present year-round, but become elusive during the breeding season – to avoid disturbance they should not be looked for at this time. Migrants from the continent may visit at any time but especially in the spring months.

■ **Habitat** Cranes feed on open, usually wet ground, such as marshland or rough pasture near open water. They may also forage in arable fields, and flyovers could appear almost anywhere.

■ **Search tips** Cranes flying overhead draw attention to themselves with their loud trumpeting calls. They fly with necks outstretched, unlike the superficially similar Grey Heron. Visiting the Norfolk Broads or the fens near Fakenham early or late in the day may produce flight views of the birds moving between feeding areas out in the fields and roosting sites within reedbeds. They usually travel and feed in small parties. In the middle of the day, scan suitable fields. The birds may be quite obvious from a distance, but on windy days tend to be more difficult to find as they hunker down in ditches or channels.

Super sites

1. RSPB Lakenheath Fen
2. Horsey Mere
3. Hickling Broad

WATCHING TIPS

These birds are shy, and because their toehold as a breeding species in Britain is so tenuous it is especially important to avoid disturbance at all costs. Good views of foraging birds can often be had quite safely from lanes or public footpaths – ideally use a telescope as the birds will probably be fairly distant. In early spring especially, you may be lucky enough to witness the courtship display performed by singles, pairs and even the whole flock together, an elegant dance where the birds strut, leap and posture with wings spread. At dusk they fly from feeding sites to a roost, often calling as they go.

Great Bustard

Otis tarda

A spectacular grassland bird, the Great Bustard became extinct in Britain in the early 19th century, but has recently been reintroduced in Wiltshire. Numbers are still very small but the early signs are encouraging

Length 75–105cm

| 1 | 2 | 3 | 4 | 5 |

J F M A M J J A S O N D

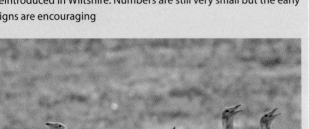

How to find

■ **Timing** Great Bustards in the UK are here throughout the year. Young birds may disperse away from the breeding grounds, meaning autumn is the likeliest time to see one away from the Salisbury Plain area.

■ **Habitat** This bird requires extensive stretches of open and undisturbed grassland, a habitat that is now very rare in Britain and becoming increasingly so on the continent. It is therefore unlikely to become a very widespread species here at any time in the near future, although grassland restoration projects could help it to spread. Wandering birds may turn up in less apparently suitable areas, such as crop fields and grazed pasture.

■ **Search tips** The sites of this species are kept secret, and it is best to avoid searching the Plain during the breeding season. At other times, be aware of the possibility of Great Bustards in fields in the general area, and try scanning fields from elevated viewpoints, being aware that these large birds can conceal themselves very effectively among long grass or crop fields. Only when performing the spring courtship display do the birds really become obvious.

WATCHING TIPS

If you should be lucky enough to find a Great Bustard, either around Salisbury Plain or elsewhere, keep a good distance away – a telescope will be helpful. The reintroduced birds have been marked with wing tags that can be read through optics at a good distance – report any sightings as the project organisers are keen to keep track of the birds' movements. Some of the tagged birds have ventured away from the Plain, turning up in Dorset, Hampshire and other nearby counties. The continued decline of the species in Continental Europe makes the survival of the new British population all the more important.

Haematopus ostralegus

Oystercatcher

An unmistakable large wader, the Oystercatcher is mainly found around our coastlines but some pairs breed well inland, especially in the north. Numbers here in winter are significantly boosted by the arrival of visitors from the continent.

Length 40–45cm

| 1 | 2 | 3 | 4 | 5 |

J F M A M J J A S O N D

How to find

■ **Timing** Oystercatchers can be seen year-round, though inland breeders move out to join the coastal breeders in winter. So not only are there more birds here in winter, but their numbers are more concentrated in a smaller area.

■ **Habitat** These birds feed mainly on mussels, cockles and similar molluscs, so need access to shallow shorelines with good supplies of these. They mainly feed on sandy or muddy estuaries, but nest on shingle shorelines or further back among dune grassland. Inland birds may nest by shingly river banks or on grass, often on lake islands. Inland, traffic islands provide them with the same safety from predators as real islands do. Unlike most waders they bring food to their chicks so a nearby food supply is not essential.

■ **Search tips** Oystercatchers are striking and often noisy, especially in spring when pairs and small flocks will fly in wide circles over their habitat, loudly piping in chorus. Scan shingly, thinly vegetated islands on coastal lakes for nesting pairs. In winter, time spent by the sea will almost certainly produce sightings of 'fly-bys' while estuaries may hold dozens of feeding birds.

WATCHING TIPS

These birds are interesting and entertaining to watch. In springtime the communal piping displays take place both in flight and on the ground, the latter involving birds rushing around with their bills pointing straight downwards. Later in the year, breeding pairs can be watched flying to and fro fetching tiny food items which they carefully feed to their chicks – this unusual behaviour enables them to nest on islands and thus stay safe from predators. Adults have two distinct ways of breaking into a mussel – by delicately prising the two halves apart, or indelicately hammering the shell open.

Avocet

Recurvirostra avosetta

The unique and elegant avian 'face' of the RSPB, the Avocet is a tremendous conservation success story. It recolonised the UK in the 1940s, beginning in East Anglia, and now breeds down the east coast of England and elsewhere.

Length 42–45cm

| 1 | 2 | 3 | 4 | 5 |

J F M A M J J A S O N D

How to find

■ **Timing** Avocets are present all year in the UK, but their numbers almost double in winter. Their distribution also changes in winter, with more northerly breeding grounds abandoned and flocks gathering along south-west coastlines.

■ **Habitat** These long-legged birds wade and swim for food, which they take from the water's surface. Fairly shallow fresh or brackish lagoons on the coast are suitable, and if these have muddy islands then they also provide breeding habitat. The birds may also feed in shallow seawater where this is sheltered and calm.

■ **Search tips** Many of the best Avocet nest-sites are on RSPB and other nature reserves, and the birds can be seen easily from hides. Bear in mind that they will happily venture into quite deep water and so may not present the expected tall and leggy outline, and scan all parts of the lagoon. Young chicks can be harder to spot – look for adults 'standing guard' on an island and the small, all-grey chicks are probably nearby. They are gregarious year-round, and in autumn and winter will roost in large, tight flocks that are obvious from a distance.

Super sites

1. RSPB Leighton Moss
2. RSPB Marshside
3. RSPB Blacktoft Sands
4. RSPB Titchwell
5. RSPB Exe Estuary
6. Poole Harbour
7. RSPB Minsmere
8. Alde/Ore Estuaries
9. Oare Marshes

WATCHING TIPS

Avocets are lively and energetic birds, and are interesting to watch with their unique, bill-sweeping feeding style. They are also very fierce in defence of their nests, and almost any passing bird may come under attack if it gets too close although potential predators like crows and heron species attract the most intense mobbing. Bear in mind that the species is still quite rare and vulnerable and take great care not to disturb nesting pairs. There are many reserves where breeding pairs and their chicks can easily be watched from hides. Winter boat trips arranged by the RSPB around the Exe Estuary give you the chance to watch large flocks.

Burhinus oedicnemus

Stone-curlew

The peculiar-looking Stone-curlew is a rare breeding species in Britain, on the far edge of its natural geographic range. It has strongholds in East Anglia and around Wiltshire. Careful management has helped its numbers increase since the turn of this century.

Length 40–44cm

| 1 | 2 | 3 | **4** | 5 |

J F M A M J J A S O N D

How to find

■ **Timing** An early returning summer visitor, the Stone-curlew is back on its breeding grounds by the end of March, with few seen away from these sites in spring. In autumn, dispersing young birds do turn up elsewhere, sometimes as late as December.
■ **Habitat** This bird is really most at home in dry heath and short grassland – inhabiting semi-desert in some parts of its range. In Britain it breeds on the Breckland heaths and on grazed downland, and on migration may visit any dry grassy, stony habitat including the vegetated upper reaches of shingle beaches.
■ **Search tips** The easiest place by far to see Stone-curlews is at the Norfolk Wildlife Trust's reserve at Weeting Heath. Here a hide allows you to view a wide area of grassland used by the birds. Even though Stone-curlews are present here through spring and summer they are still not necessarily easy to spot, as they are rather inactive in the day and sit low in the grass. Heat haze can also be a problem. However, with patience you should find them. Other breeding sites are not publicised. In suitable habitat listen for the haunting, curlew-like call at dusk.

Super sites

1. Weeting Heath

WATCHING TIPS

The Stone-curlew's semi-nocturnal habits make it a difficult bird to study. Early or late visits to Weeting Heath (the reserve is open from 7am to dusk in spring and summer) are best for seeing some activity, especially early in the breeding season when the grass is short and the birds may be engaged in courtship behaviour. This is remarkably showy – the birds rush about in pairs with wings and tail fanned, giving wheezing throaty calls. Should you encounter a Stone-curlew away from this site in the breeding season, keep a good distance away and inform the local bird recorder.

Little Ringed Plover

Charadrius dubius

A recent UK colonist, the Little Ringed Plover's first breeding record was in Hertfordshire in 1938, and it has since spread north and west. Its range now includes eastern Wales, though it is yet to reach Scotland and Ireland.

Length 14–15cm

| 1 | 2 | 3 | 4 | 5 |

J F M A M J J A S O N D

How to find

■ **Timing** This plover is one of the earliest of summer visitors, with arrivals in February not exceptional. It is also early to depart, with only stragglers remaining into September. They are very visible and active in spring but become more discreet in summer.

■ **Habitat** Most pairs nest inland on the shorelines or (more usually) islands of inland reservoirs and gravel pits, or permanent gravel banks within wide rivers. A shingly substrate is preferred, providing good camouflage for the eggs. Newly arrived birds and, later in the year, dispersing youngsters visit other wetland sites including coastal lagoons with shingle or gravel shorelines.

■ **Search tips** Although small and well camouflaged, the Little Ringed Plover often draws attention to itself through activity. If several birds are present at a site you will probably see them chasing each other around on the islands or in a low, fast flight. If the birds are not active, careful scanning of shorelines and islands should prove successful. To be sure you are looking at Little Ringed and not Ringed Plovers, check leg colour and presence or absence of wing-bars. The Little Ringed Plover's yellow eye-ring and slimmer shape is also often obvious even from a distance.

WATCHING TIPS

In early spring, these lively birds provide entertainment at a time which is generally rather quiet for bird life at inland waters. The territorial chases show off their remarkably fast and agile flight. Once paired up and nesting, the sitting bird is hard to spot but its mate will patrol nearby, launching an aggressive attack on any passing bird or other animal that it perceives to be a threat. Juvenile Little Ringed Plovers can turn up in surprising places, perhaps rubbing shoulders with young Ringed Plovers and providing an identification challenge.

Charadrius hiaticula

Ringed Plover

This common and attractive plover is one of the few British birds adapted to nest on shingle beaches, so its distribution is mainly coastal, but it also breeds inland, especially in the north. In winter numbers increase substantially.

Length 18–20cm

| 1 | 2 | 3 | 4 | 5 |

J F M A M J J A S O N D

How to find

■ **Timing** Ringed Plovers are easier to see in winter but are fairly easy to find at any time of year. Be very careful when walking across beaches used by nesting birds in spring, as their eggs are camouflaged and so could easily be accidentally destroyed.

■ **Habitat** Nesting habitat is on shingly or gravelly ground, which may be on the coast or inland, alongside rivers or gravel pits. Pairs that choose islands, inaccessible to mammalian predators, are more likely to be successful. However, any 'ringed plovers' seen at inland gravel pits in England are more likely to be Little Ringed Plovers. Foraging birds will use sandy or muddy seashores and shorelines of coastal lakes and estuaries, especially in winter.

■ **Search tips** This boldly marked bird can be surprisingly hard to spot against a backdrop of pebbles. Getting a low perspective, so the bird's outline breaks the skyline, can help. In winter, look out for Ringed Plovers among other wading birds on the seashore or in mobile flocks. Their activity pattern is noticeably different from other small waders from a distance – rather than being constantly active they stand still for spells, in between making short dashing runs.

WATCHING TIPS

Because these plovers rely heavily on camouflage (of adult, egg and young chick) to protect their nests, entering their nesting habitat should be avoided if possible. The adult may perform a 'distraction display' intended to lure you away from the nest or chicks – this involves feigning injury and, while interesting to see, is a sure sign that you are too close for comfort and should back off. In winter, finding a spot on the beach as the tide is rising and waiting quietly could give you close views as the birds are pushed up the beach by the incoming water.

Kentish Plover

Charadrius alexandrinus

The Kentish Plover, as its name suggests, used to breed in south-eastern England, but now only occurs in the UK as a scarce visitor, though still with a south-easterly bias. It may be confused with Ringed and Little Ringed Plovers.

Length 15–17cm

1	2	3	4	5

J F M A M J J A S O N D

1&2
3
4&5

How to find

■ **Timing** This species is a summer visitor to northern Europe. It may turn up in the UK any time in spring or summer, though May is probably the best month, with returning adults overshooting their breeding grounds. Late summer is also a good time, as juveniles disperse before migrating south.

■ **Habitat** Kentish Plovers are primarily birds of the seashore and of estuaries and brackish coastal lagoons. Sandy shorelines seem to attract them most. Inland records are uncommon. Juveniles are often found associating with juvenile Ringed Plovers or other waders, on quiet beaches along the south and east coasts.

■ **Search tips** As with other overshooting migrants, adult Kentish Plovers are most likely to arrive in the UK during spells of warm, fine weather further south, especially if there are also southerly winds. Scan suitable beaches and check any small plovers, looking out for any that look particularly pale and/or 'front-heavy'. These migrants may be long-stayers, perhaps prospecting for potential new breeding sites. In late summer and early autumn, carefully check through all groups of Ringed Plovers and other shoreline waders.

Super sites

1. RSPB Cliffe Pools
2. Shellness Beach on Sheppey
3. Sandwich/Pegwell Bay
4. Rye Harbour
5. RSPB Dungeness

WATCHING TIPS

This bird has not long been lost as a British breeding species, and could well become re-established here in the future. Birds found in spring should be considered to be potential breeders and given plenty of space to avoid disturbance. Watching from a good distance, you should be able to appreciate the differences between this species and the Ringed Plovers that are probably also present. In breeding plumage the two are quite distinct, but juveniles are more similar and this is when the subtle differences in proportions become especially important for identification.

Charadrius morinellus

Dotterel

A beautiful plover of the uplands, the Dotterel is a rare and very localised breeding bird in Britain, restricted to Scottish mountains. Migrating birds stop off at favourite spots on their northward journey – autumn birds are less predictable.

Length 20–22cm

| 1 | 2 | 3 | **4** | 5 |

J F M A M J J A S O N D

How to find

■ **Timing** In April and early May, parties or 'trips' of Dotterels may be seen in open country almost anywhere in the UK, although there are a few traditional sites that attract birds every year. These stop-offs may last several days. By mid-May they will be on their breeding grounds.

■ **Habitat** Dotterels nest on high open ground, with thin vegetation and rocky ground. Most pairs are found higher than 1,100 metres above sea level. On migration, they favour open grassland or arable fields. There is no marked preference for high ground – autumn birds in particular may spend time at or near sea level, often in company with Golden Plovers.

■ **Search tips** Making regular visits to one of the traditional sites in spring should produce sightings. The birds are not especially conspicuous on the ground so scan carefully – the same applies when looking for them in the mountains in summer (keep to footpaths to avoid the risk of disturbance or accidental damage to a nest). In autumn, anywhere that attracts post-breeding flocks of Golden Plovers is worth a visit – check through the plover flocks carefully.

Super sites

1. Carn Ban Mor
2. Cairngorms
3. Pendle Hill
4. Burbage Moor
5. Gringley Carr
6. Happisburgh

WATCHING TIPS

Dotterels are noted for being almost fearless of humans, so when watching them keep still and quiet and let them come to you and you may be lucky with some extremely close views. However, it's important to keep to footpaths and avoid walking across sensitive habitat – bear in mind at all times that any form of disturbance to either migrating or breeding must be avoided. Spring birds have often already formed pairs – note the interesting reversed sexual dimorphism the species shows, with the females larger and brighter than their mates.

Golden Plover

Pluvialis apricaria

The Golden Plover breeds in mainly northern uplands, and in winter moves south and to lower ground – depending on time of year it may be found anywhere in Britain. Extra birds arrive from the continent through the winter months.

Length 26–29cm

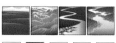

1	2	3	4	5

J F M A M J J A S O N D

How to find

■ **Timing** This species returns to its moorland breeding grounds in early May, and most birds return south through September and October. Numbers peak in mid-winter, with the best locations hosting very large flocks.

■ **Habitat** Breeding habitat is typically open heather moorland, between 240 and 600 metres above sea level – so not usually reaching into truly mountainous terrain. In winter, Golden Plovers flock on low and flat arable and grazed fields, often adjoining water or on the coast and sometimes on the upper slopes of undisturbed beaches, but rarely down on the seashore. They frequently team up with flocks of Lapwings.

■ **Search tips** Like other plovers, this species is a ground-nester and is well camouflaged in its breeding habitat. Search carefully and slowly, and be careful not to wander across moorland, to avoid disturbance or damage to nests. From autumn, as flocks start to form, check suitable fields and shores and islands of shallow lakes, especially if you notice Lapwings around, and listen for the distinctive fluty flight call. Flocks on the wing are fast-flying and agile with eye-catching flickering wingbeats.

WATCHING TIPS

When they first return to the nesting grounds, male Golden Plovers perform a songflight over their territories, flying with slow graceful wingbeats to show off their bright white 'armpits' to any nearby rivals. This may be seen later in the breeding season, but a visit to the moors in May is your best bet to witness it. When a mixed winter flock of Golden Plovers and Lapwings is flushed by a raptor, it is interesting to note how the Golden Plovers quickly segregate themselves into a tight flock, which flies in wide circles and usually stays airborne for longer.

Grey Plover

A stocky plover, this species does not breed in Britain but visits in winter and on migration, and a few non-breeding birds are present through the summer. Its distribution in Britain spans all the countries but is almost exclusively coastal.

Length 27–30cm

| 1 | **2** | 3 | 4 | 5 |

J F M A M J J A S O N D

How to find

■ **Timing** Some birds linger well into May before migrating north-west to their Arctic breeding grounds, and good numbers of passage migrants are already here by the end of August. Some of these migrants will move on but numbers through winter are not very much lower than during the passage periods.

■ **Habitat** Grey Plovers like muddy and sandy coasts, including sandy expanses at the foot of shingle beaches that are only exposed at low tide. They will also feed on low shoreline rocks. It is unusual to observe them inland, though they may visit the shores of coastal lagoons.

■ **Search tips** In winter plumage, the Grey Plover is a rather undistinguished-looking bird and stands out among flocks of shoreline waders more by behaviour than appearance. Look for a largish, solid-looking wader, often on its own, that stands stock-still for long periods in between short, rapid dashes. Though the bill is short for a wader it often looks rather broad and heavy, with a 'swollen' tip. In flight, the jet-black axillaries ('armpits') are striking.

WATCHING TIPS

Much less 'busy' than most of the other waders, Grey Plovers can be quite confiding and if you wait patiently in an inconspicuous viewpoint, you could be lucky enough to enjoy close views. In spring and autumn, many of the migrants that stop off on our coasts will be in their very attractive full breeding plumage. Individual Grey Plovers may establish temporary feeding territories on the shore and drive off others, though as the rising tide restricts available foraging ground they become more tolerant of each other. It may be confused with Golden Plover – look for the hefty-looking bill and greyer plumage.

FOUND A RARE BIRD?

Most of us hope to find something out of the ordinary when we are out birdwatching, however much we enjoy watching the commoner species. Finding a rarity is a real thrill, and if you choose to pass on the news then many other birdwatchers may be able to enjoy the sighting. If you should find rare breeding birds, a more cautious approach is needed to ensure they are not unwittingly disturbed or deliberately targeted by egg collectors.

Are you sure?

Sometimes a rarity is easy to identify. You are unlikely to confuse a Hoopoe or Bee-eater with any other species. Other rarer birds may be less familiar, but stand out so much as 'something different' that you will certainly notice them, and be able to identify them easily when you consult your field guide.

The majority of species, though, are not quite so simple to identify, and can be easily confused with at least one other more common species. Your first task is to eliminate those confusion species from your enquiries, and this will be done more quickly and easily if you are very familiar with them. So if you want to find a rarity, it helps a great deal if you have 'paid your dues' and spent plenty of time studying the common species.

If you are still not sure, hopefully there will be other birdwatchers around who can help you. If not, take photos if you have a camera (even poor photos can be enough to confirm an identification), or take notes and make sketches. You can then consult a field guide and also discuss the identification with other birdwatchers, via online forums if not in person.

Spreading the word

Once you are confident that you have found a rarity, you need to decide whether to let other birdwatchers know. In most cases, there are no problems with passing on the news, but there are a few circumstances where you may want to think twice.

■ The bird is on private land. If this is the case, it may not be possible to arrange access for visiting birdwatchers. Talk to the landowner before releasing any news.

■ The bird is in an area where other resident breeding or wintering birds, or other wildlife, would suffer serious disturbance if large numbers of birdwatchers visited. The welfare of all birds and the environment must come first.

■ You suspect the bird may be breeding. In these circumstances, the sighting should be passed to your county recorder only.

When many birdwatchers come to see a rarity, there is always a small risk of problems, but most of these 'twitches' go smoothly. Wildlife-friendly landowners sometimes use the occasion as an opportunity to raise money for conservation charities, by accepting a small donation in exchange for access.

Red-breasted Goose with Greylags

Ring-billed Gull with Black-headed Gulls

RARE BREEDING BIRDS

Many of the locations where rare birds are known to breed are not publicised in any way. Although birdwatchers would love to see them, it is more important to ensure that disturbance is kept to a minimum, to give them the best chance of breeding successfully. This means it is quite feasible that you could stumble upon nesting rarities while walking on ordinary public footpaths in the countryside – when you pass on details of your sighting to your county recorder, they will let you know whether you've found a known nest-site or a 'new' one.

SUBMITTING A SIGHTING

The BTO's website has a full listing of the current county recorders – the people you should inform if you find a rarity. Most can be contacted by email. When you submit your sighting, include the following information:
■ Date and time of day of the sighting
■ Location (supply a six-figure grid reference if possible)
■ Number of birds seen
■ Direction of travel if the bird/s were in flight
■ Any interesting behaviour noted (especially signs of breeding behaviour)
■ You should also attach any photos you took, or field sketches and field notes. The notes should include details of how you were able to rule out common confusion species.

It is not essential to submit your sightings. However, by doing so you are helping to build up a picture of bird distribution and migration patterns, and if the birds in question are breeding, your report could make the difference between their success and failure.

Golden Oriole

Lapwing

Vanellus vanellus

Probably our most distinctive wader, the Lapwing is common on farmland and wetlands throughout Britain, despite having declined recently. In winter its numbers increase dramatically as birds from the continent visit, drawn by our milder weather.

Length 28–31cm

1	2	3	4	5

J F M A M J J A S O N D

How to find

■ **Timing** Lapwings may be seen year-round. Large concentrations are present in winter, while the wonderful territorial display flight can be seen through the breeding season but especially in early spring. The charming chicks hatch in late spring.

■ **Habitat** For breeding, this species requires flat and usually low-lying open countryside with enough vegetation cover to conceal a nest and small chicks, and access to soft ground with a rich population of invertebrates. Pasture and arable fields may both be used, although sympathetic management is necessary for successful nesting on farmland. In winter similar habitats are used, along with the shores and islands of lakes and reservoirs, both inland and by the coast.

■ **Search tips** With colourful and boldly contrasting plumage along with an extrovert character, the Lapwing is easy to see. Look out for single birds (in spring) or flocks (in winter) wheeling over farm fields, the very broad black and white wings are eye-catching from a distance. When birdwatching at any lowland lake, scan islands and along the shore for resting Lapwings.

WATCHING TIPS

Lapwings are lively birds and very entertaining to watch. Even in winter when they spend time resting, conserving their energy, they will quickly rise up to mob a predator or other perceived threat. In the breeding season, both members of the pair will circle the territory, calling loudly and performing dramatic steep tumbles in the air, in between driving off other birds. By mid-spring tiny chicks may be seen exploring the grassland, before hurrying back to brood under their mother's wings, while the male surveys the scene and takes off to attack any passing gull, crow or heron.

Knot

This chunky medium-sized wader is a passage migrant and winter visitor to Britain, and occurs in very large concentrations at the major estuaries all around Britain. Away from these key areas, though, it can be surprisingly scarce.

Length 23–25cm

| 1 | **2** | 3 | 4 | 5 |

J F M A M J J A S O N D

How to find

■ **Timing** Knots start to arrive in Britain as early as August, and can be seen right through until May. The numbers are highest between December and February, and you will get better views by visiting estuaries at high tide, especially the highest 'spring' tides.

■ **Habitat** These waders feed on sandy or muddy shorelines exposed by the retreating tide, on the coast or in the mouths of estuaries. When very high tides cover up all of the feeding ground, the Knots will roost in tight flocks higher up on the beach, or move to safe roosting grounds on, for example, islands in coastal lagoons. This species is very rare inland but is occasionally found on lake or reservoir shorelines.

■ **Search tips** At its key sites, the very gregarious Knot is often by far the most numerous wader and can be seen in flocks thousands strong. Flocks in the air are typically very tight and co-ordinated, and at high-tide roosts they will pack close together, carpeting the ground. Singles or smaller groups may be found among other waders, feeding on the shore, and stand out with their bulky build and more deliberate feeding movements compared to smaller shoreline waders.

Super sites

1. RSPB Nigg Bay
2. RSPB Hodbarrow
3. RSPB Morecambe Bay
4. RSPB Dee Estuary – Point of Ayr
5. RSPB Frampton Marsh
6. RSPB Snettisham
7. RSPB Stour Estuary

WATCHING TIPS

If visiting a high-tide roost site used by Knots, try to arrive an hour or two before high tide and make sure you are in a spot where you will not disturb the birds as they seek refuge from the rising water. From a hide you should enjoy great views of the flocks. If disturbed by a raptor, the birds will take off en masse and perform extraordinary aerial manoeuvres, twisting and turning as if of one mind to try to confuse the predator. The earliest arriving (and latest departing) flocks may include some birds sporting their colourful breeding plumage.

Sanderling

Calidris alba

A charming small wader, the hyperactive little Sanderling is primarily a passage and winter visitor to Britain, although a few non-breeders linger through the summer. It can be found on suitable shorelines anywhere around the coast.

Length 20–21cm

| 1 | 2 | 3 | 4 | 5 |

J F M A M J J A S O N D

How to find

■ **Timing** Sanderling numbers in the UK begin to build up through September. Wintering birds remain until mid-spring, when they are joined temporarily by returning migrants that wintered elsewhere. Only a handful of non-breeders oversummer here, but migrants are present until late spring and then again from late summer.

■ **Habitat** The name of this bird gives the clue to where it is usually found – flat, sandy beaches. It shows a preference for sandier rather than muddy beaches compared to most waders, and so may be numerous on beaches where other waders are scarce or absent. It is uncommon inland, records usually relating to solitary stragglers.

■ **Search tips** Look out for Sanderlings close to the water's edge, in small groups. Their movements are lively and restless, rushing close to the incoming waves and away again, and often taking off and flying short distances in tight, flickering flocks. At all times, a clear view of a Sanderling will reveal its key identifying trait compared to all other small waders – the lack of a hind toe.

WATCHING TIPS

Sit high on the beach as the tide is rising and with luck you will see Sanderlings at close range as the incoming water pushes them towards you. Their frenetic feeding behaviour is entertaining to watch, and if you keep still and quiet they will often show little fear. In mid-winter they look strikingly pale, their plumage mostly white with a blackish patch at the bend of the wing, but at migration times they may show some or much reddish coloration, with one flock exhibiting a challenging range of different plumages. Check for other species within the flock when watching feeding Sanderlings.

Calidris minuta

Little Stint

The smallest wader seen regularly in Britain, the Little Stint does not breed or overwinter in Britain, but is a passage migrant mainly to eastern coastlines, with smaller numbers visiting more westerly coasts and the shores of inland waters.

Length 13–15cm

| 1 | 2 | 3 | 4 | 5 |

J F M A M J J A S O N D

How to find

■ **Timing** The first Little Stints of autumn appear in early August. Passage continues through early autumn, and numbers are dwindling by mid-October. One or two may stay on for winter. This species is much more numerous on autumn migration than in spring, but there is a small return passage between late April and early June.

■ **Habitat** Sheltered muddy shorelines and estuaries attract Little Stints, both on beaches and around coastal lagoons and ponds. Inland, they may use the shores of lakes and reservoirs, especially if summer drought reveals plenty of mud, and riversides.

■ **Search tips** This species stands out among other waders because of its tiny size. Any time you encounter a mixed group of shoreline waders, look carefully through them, bearing in mind that Little Stints can easily be hidden among larger birds or concealed by a subtle undulation in the ground. They often associate with Dunlins, and small Dunlins may be close to Little Stint size and proportions so careful study is necessary. Look around even the smallest pools and ditches as well as more likely-looking spots.

Super sites

1. RSPB Saltholme
2. RSPB Dee Estuary – Parkgate
3. RSPB Frampton Marsh
4. RSPB Titchwell
5. Slimbridge WWT
6. RSPB Bowling Green Marsh
7. Abberton Reservoir
8. RSPB Elmley Marshes
9. Oare Marshes
10. Weir Wood Reservoir
11. Sandwich Bay

WATCHING TIPS

The size of this bird can be quite startling, especially if seen at close quarters. Barely bigger than a sparrow, the beautiful details of its plumage are hard to appreciate from any distance. Watching from a hide gives you the best chance of a good view and to compare it to similar species. Most Little Stints in Britain are juveniles, and as such can be very confiding when encountered in the open, but remember that they are in the middle of an arduous migration and allow them plenty of space.

Temminck's Stint

Calidris temminckii

This stint is a passage migrant to the south-east, commoner in spring, and also a very rare breeding bird in Scotland.

Length 13–15cm

| 1 | 2 | 3 | **4** | 5 |

J F M A M J J A S O N D

How to find it

■ **Timing** May is the best month for Temminck's Stints in Britain.
■ **Habitat** It prefers fresh water to salt, and turns up most often on the shores of coastal freshwater marshes and estuarine pools and creeks, but rarely the seashore.
■ **Search tips** Scan patches of exposed mud in suitable marshes, looking for a tiny, low-set, deliberately moving wader. Avoid searching breeding habitat in summer, as the breeding population here is tiny and precarious, and no sites are currently set up for visitor access.
■ **Super sites 1.** RSPB Saltholme, **2.** RSPB Blacktoft Sands, **3.** RSPB Titchwell, **4.** Cley, **5.** RSPB Buckenham Marshes, **6.** Grove Ferry, **7.** RSPB Dungeness.

WATCHING TIPS

Because of their differing habitat needs and migration patterns, Temminck's Stints are not often confused with Little Stints. However, besides these two species there are several extremely rare stints, and it is a good idea to take notes and/or photos to confirm identification.

Pectoral Sandpiper

Calidris melanotos

This wader is a rare visitor from North America, most likely to be found on the west coast in autumn.

Length 22–23cm

| 1 | 2 | 3 | 4 | **5** |

J F M A M J J A S O N D

How to find it

■ **Timing** From late August through autumn, Pectoral Sandpipers turn up throughout Britain, with arrivals most likely following a spell of strong westerly winds.
■ **Habitat** Fresh or saltwater marshland with exposed mud and shoreline vegetation is the likeliest habitat. It rarely visits the seashore.
■ **Search tips** Pick a suitable day (weather wise) and site, and scan marshland edges. This bird is superficially similar to the Dunlin, so check any unusual-looking large Dunlins.

WATCHING TIPS

After severe westerly gales, this and other North American waders can appear in double figures at well-placed sites. Take notes or photos to confirm identification. Spring records are rare but may relate to birds that arrived the previous autumn and successfully overwintered in Europe.

Curlew Sandpiper

A small but rather graceful and long-billed shorebird that breeds in the Arctic, the Curlew Sandpiper visits mainly marshes and estuaries around most of the British coast on its autumn migration.

Length 19–21cm

1	2	**3**	4	5

J F M A M J J A S O N D

How to find

■ **Timing** The first returning migrants – typically adults that failed to breed – appear as early as late July. The main influx, mostly juveniles, is through September into October. Returning spring migrants are very rare.

■ **Habitat** Although its distribution is very coastal, this is another wader that usually avoids the open seashore and is more likely to be found in shallow, mud-fringed coastal floods, estuaries and lagoons, often in company with Dunlins and other waders.

■ **Search tips** Check the shores, islands and shallows of suitable marshes and estuaries, looking for a bird that stands somewhat taller than the Dunlins that are almost certainly also present. If the birds are sitting down, look for the Curlew Sandpiper's longer bill and more strongly marked face. Juveniles have a distinctive and beautiful pale peachy wash to the plumage, and early adults may show plenty of the bright brick-red breeding plumage, but winter-plumaged adults are quite dull, and look rather pale alongside Dunlins. As with most waders, the flight call is distinctive, and when in flight the neat, square, white rump patch is striking.

Super sites

1. RSPB Belfast Lough
2. Banks Marsh
3. RSPB Dee Estuary – Burton Mere Wetlands
4. Slimbridge WWT
5. RSPB Titchwell
6. RSPB Breydon Water
7. RSPB Cliffe Pools
8. Oare Marshes
9. Pett Pools

WATCHING TIPS

These waders, though often confused with Dunlins, have a demeanour and elegance that recalls a larger wader. When actively feeding, they will readily wade belly-deep, catching small prey items from the water's surface. Visit suitable sites early in the season for the best chance of seeing them in their spectacular breeding plumage. Much of the day may be spent in relative inactivity, with early mornings and evenings best for watching the birds in action.

Purple Sandpiper

Calidris maritima

This winter-visiting wader is strictly coastal, and is most numerous on the east coast of Scotland and northern England, but it occurs patchily elsewhere. It is also an extremely rare breeding bird in the Scottish Highlands.

Length 21–23cm

1	2	3	4	5

J F M A M J J A S O N D

1,2&3

4

6

How to find

■ **Timing** Purple Sandpipers arrive from mid-autumn, but the middle of winter is the best season for them. They linger here until mid-spring before commencing their return migration to the Arctic.

■ **Habitat** This species is very rarely seen away from the sea coast, and shows a preference for rugged and rocky shorelines. It frequently roosts on or forages along man-made structures such as groynes, piers and marinas that project into the sea, often in company with Turnstones. It may also forage along undisturbed shingle beaches, but tends to avoid expanses of flat sand or mud.

■ **Search tips** Not purple but dark grey, this sandpiper is well camouflaged and easily overlooked as it moves over dark rocks or roosts among them. Scan rocks, ledges and other suitable projections right at the sea edge, and when walking down piers or around harbours take a look over the edge at where the structure meets the water. Check any Turnstone flocks you find, as single Purple Sandpipers will often join them. Avoid searching for breeding Purple Sandpipers as they are extremely rare and vulnerable.

Super sites

1. Scarborough
2. Filey Brigg
3. Bridlington
4. Sennen Cove
5. Penzance
6. Brighton Marina

WATCHING TIPS

These waders are oddities among sandpipers, moving rather deliberately over rocks when feeding, and picking molluscs from the rocks and small crustaceans from rock pools. Like the Turnstones with which they so often associate, they can be very confiding, especially where they share their beach with a steady stream of human visitors. Notice which way they are moving and wait in a position ahead of them, so their foraging path will take them closer to you, with very close views, you may be able to make out the vague purplish lustre to the plumage gives the species its name..

Calidris alpina

Dunlin

The Dunlin is generally our commonest small wader, and the yardstick to which the others are compared. It can be found along most coastlines in winter, and breeds in the northern uplands, especially in Scotland and north-west Ireland.

Length 17–21cm

| 1 | 2 | 3 | 4 | 5 |

J F M A M J J A S O N D

How to find

■ **Timing** Dunlins may be found along the coast all year round, though are scarce or absent on southern coastlines in summer. They occupy their breeding grounds between April and June.

■ **Habitat** At the coast, they feed on muddy and sandy shorelines up to the edge of the sea, but will also use estuarine mud, creeks and pools, and lagoons and floods near the coast. When high tide covers the beach they may retreat to fields just inland. They breed in undisturbed open uplands with grassy or heathy moorland.

■ **Search tips** Most gatherings of waders at or by the coast will include some Dunlins. Unless Little Stints are also there, the Dunlins will be the smallest birds present, and in spring and autumn many will show at least some signs of their black belly patch, a feature not found on other waders. The feeding behaviour is active yet methodical, the bird picking rapidly at the ground as it runs. Breeding birds call and display in the air at the start of the season, but become very discreet once nesting is under way.

WATCHING TIPS

It is well worth becoming familiar with the Dunlin in order to be able to rule it out quickly when you find what you think may be a rare small wader. It is a variable species, with plumage intergrades along with different subspecies to add to the confusion. Within one flock you may notice considerable individual variation in body size, plumage tones and bill length. Dunlins in breeding plumage are very attractive birds, and may be seen around northern coasts in spring as well as in their breeding areas. Winter flocks on the move demonstrate remarkable coordinated twists and turns, especially when trying to evade or confuse a bird of prey.

Ruff

Philomachus pugnax

Although some Ruffs are in Britain all year, numbers are highest in the migration periods and the south and east coasts are the best areas. There is a small and apparently dwindling breeding population in Eastern England.

Length 22–32cm

1	2	3	4	5

J F M A M J J A S O N D

How to find

■ **Timing** Migrating juvenile Ruffs visit Britain from mid-summer, while adults that arrive later in autumn will often remain through the winter. Early summer is the time when fewest Ruffs are present, though this is when the breeding birds will be active.

■ **Habitat** Ruffs feed on both salt and freshwater marshes, which may be well vegetated or muddy. They will also use the margins and islands of reservoirs and lakes inland, and may forage on wet grassland. The breeding habitat is quiet low-lying grassland.

■ **Search tips** This is the wader that catches out many beginner birdwatchers, as it is so variable in size and plumage. In late spring and late summer you may see unmistakable adult males with intact 'ruffs' on their heads, but females and young birds are rather dull. At all times, look for a medium-sized wader with a proportionately small head and shortish bill, and scaly-looking plumage on the back. Check grassy islands, shorelines and also open water as they will wade quite deeply, and check for 'hangers on' around the edges of flocks of other waders such as Black-tailed Godwits.

Super sites

1. Rutland Water
2. RSPB Titchwell
3. RSPB Ouse Washes
4. RSPB Minsmere
5. RSPB Cliffe Pools
6. Oare Marshes
7. RSPB Dungeness

WATCHING TIPS

It has been said that there is a 'Ruff for every day of the year', and watching this species in its many guises is worthwhile for anyone with an interest in waders. The warm brown juveniles bear an uncanny resemblance to the rare Buff-breasted Sandpiper, adults are commonly mistaken for Redshanks, and males with traces of their breeding finery can look simply bizarre. The Ruff is a lekking species, the colourful males displaying communally to the females, but its rarity as a British breeding bird means that only a lucky few have witnessed the display here.

Lymnocryptes minimus

Jack Snipe

This extremely skulking small wader is a winter visitor to much of Britain, only absent from the uplands. It is very difficult to see, and frequently confused with the Common Snipe, although there are reliable ways to tell the two apart.

Length 18–20cm

| 1 | **2** | 3 | 4 | 5 |

J F M A M J J A S O N D

How to find

■ **Timing** Jack Snipes are present in Britain between mid-autumn and early spring. They tend to be most active around dawn and dusk.
■ **Habitat** This is a species of wet marshland, with dense vegetation and small muddy pools or creeks with soft ground where it can probe for food. It may also feed along the muddy edges of reedbeds. Freezing weather may drive them into less optimal habitat.
■ **Search tips** Most Jack Snipes are seen by accident, flushed from their hiding places when you walk too close. They will wait until the last moment before breaking cover, and will skim low over the vegetation, quickly dropping back down again. Common Snipes have different flushing behaviour, making a zigzag escape flight and also often calling when flushed, while the Jack Snipe stays silent. For a more rewarding view of this species, you may need a patient wait in a hide that gives views of suitably muddy but well vegetated shores. A freeze-up may force them into more open areas. Sites that hold good numbers of Common Snipes in winter are often also home to a few Jack Snipes.

Super sites

1. RSPB Saltholme
2. RSPB Old Moor
3. Spurn Head
4. RSPB Titchwell
5. RSPB Strumpshaw Fen
6. St Mary's Island
7. RSPB Rainham Marshes

WATCHING TIPS

It is worth taking the time to find a place where you stand a chance of watching a Jack Snipe without disturbance, as they are endearing and characterful little birds, with a distinctive bobbing 'clockwork toy' motion when feeding. Good views will reveal the distinct head pattern with its curious 'eyebrow' marking that helps distinguish the species from Common Snipe. When seen side-by-side, the two species show rather little difference in size. The very cryptic plumage provides superb camouflage but is beautifully intricate when examined closely.

Common Snipe

Gallinago gallinago

Although quite a common and very widespread species, its shy habits mean that the Common Snipe is not always easy to observe. It may be seen year-round in most areas; the breeding population is most dense in northern and upland areas.

Length 25–27cm

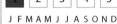

| 1 | 2 | 3 | 4 | 5 |

J F M A M J J A S O N D

How to find

■ **Timing** To see and hear displaying Common Snipes on their breeding grounds, visit between April and July. Wintering flocks build up through autumn to peak in mid-winter, and freezing weather forces them to disperse to more marginal habitat. Most activity is around dawn and dusk.

■ **Habitat** Common Snipes breed on wet grassland and moorland, with plenty of ground cover to hide a nest, and wet boggy places for feeding. Away from the breeding grounds, well-vegetated marshes, wet pasture with ditches, reedbed edges and muddy lagoons offer suitable feeding ground. In severe winter weather Common Snipes occasionally turn up in gardens.

■ **Search tips** Singing males produce a very distinctive bleating or kazoo-like 'drumming' sound (made by the vibrating tail feathers) as they perform their display flights – hearing this sound is the surest way to locate a Common Snipe on its breeding grounds. When walking in suitable habitat, you may flush a bird close to your feet, which will give a croaking call as it zigzags away. On marshland, scan the shore and island edges for a slowly moving, chunky and heavily streaked bird.

WATCHING TIPS

The unearthly sound of a drumming Common Snipe is almost unnerving if you aren't expecting it, but this is your chance to watch the bird's display flight. You may even spot the motion of its tail feathers that produces the sound. Outside the breeding season, Common Snipes often become more adventurous and feed more in the open towards dusk. In winter, feeding birds may assemble in small parties as sunset approaches, their plumage showing to its best advantage in the evening light. Always check parties of Common Snipes in winter in case there is a Jack Snipe among them.

Woodcock

A shy woodland bird, the beautiful Woodcock occurs throughout most of the UK and Ireland year-round, although it is scarce in the uplands in winter, and is absent from most of south-west England in summer. It is easiest to see on summer evenings.

Length 33–38cm

1	2	3	4	5

J F M A M J J A S O N D

How to find

■ **Timing** Woodcocks tend to be very difficult to see, with two exceptions – in early summer the males perform a very visible and vocal aerial display, and in severe winter weather they may be forced out of their preferred sheltered woodlands into more open locations.

■ **Habitat** Damp woodland suits these birds, which feed by probing soft ground with their very long bills. Both coniferous and deciduous woodland may be used, especially woods with little human disturbance. Cold weather on the continent brings migrants here, which may be found in surprising places (even city centres), and in a freeze birds will disperse to anywhere with soft ground.

■ **Search tips** By far the easiest way to see a Woodcock is to visit a suitable woodland as dusk is approaching and watch and listen for a male performing its territorial display. Listen for a single squeaking or grunting call given at regular intervals, and scan the sky for a chunky bird with pointed wings, flying in straight lines quite high over the trees. In winter, a patient vigil from a well-hidden spot overlooking a woodland pond with muddy margins may produce a sighting.

WATCHING TIPS

Roding Woodcocks will often give clear, close and prolonged views, albeit in challenging light conditions. Use binoculars with large objective lenses as these will gather the most light, and find a spot that the bird is overflying on its circuits – males tend to stick to the same general flight paths as they patrol, and are unworried by the presence of a human on the ground. If you find a Woodcock in severe winter weather, especially away from suitable wooded, marshy habitat, check that it is alert and looks well – some such birds are exhausted and need to be taken into care and rehabilitated.

Black-tailed Godwit

Limosa limosa

This very stately and elegant large wader is primarily a passage migrant and winter visitor to Britain, mainly to the coasts of England, south Wales and Ireland. East Anglia has a small breeding population, as does Shetland.

Length 37–44cm

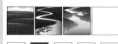

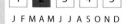

| 1 | 2 | 3 | 4 | 5 |

J F M A M J J A S O N D

How to find

■ **Timing** Migrants begin to arrive on coastal wetlands from late summer, including the East Anglian and other mainland European breeding birds, and those from Iceland and Shetland, which are of a different subspecies. The Icelandic birds overwinter in Britain, not departing until mid-spring, while the European birds continue to migrate further south.

■ **Habitat** Coastal marshes with stretches of shallow water for feeding and islands for roosting attract Black-tailed Godwits. They will also feed and roost on wet pasture. The species prefers fresh water, a point which helps distinguish it from the Bar-tailed Godwit. Breeding habitat is wet, open grassland, as found in the fenlands of west Norfolk and Cambridgeshire.

■ **Search tips** These large waders are usually quite obvious, feeding in flocks belly-deep in the marshland waters. When not feeding they will sit in large, dense flocks on islands or in shallow water. Look for tall, leggy birds with long, pink- or orangey-based bills. In flight the white rump and broad white wing-bars are striking, as are the long bill and legs.

Super sites

1. Shannon Estuary
2. RSPB Dee Estuary
3. The Wash
4. RSPB Titchwell
5. RSPB Ouse Washes
6. RSPB Minsmere
7. RSPB Stour Estuary
8. Oare Marshes

WATCHING TIPS

The two different subspecies of Black-tailed Godwits that visit Britain are quite distinct, and in spring you can see them side by side in their lovely red breeding plumage. The Icelandic form is smaller, shorter-billed and more colourful than its European counterpart. If you visit the Ouse Washes area in early summer you may be lucky enough to witness the territorial song flight. Feeding Black-tailed Godwits are very active, energetically dipping their bills and sometimes heads and necks into the water. They are also highly gregarious but tend not to form mixed flocks with other species.

Limosa lapponica

Bar-tailed Godwit

This godwit is a winter visitor to Britain, and may be found on almost any stretch of suitable coastline, although numbers are highest in and around the major estuaries. With good views it is easy to tell from the Black-tailed Godwit.

Length 37–41cm

| 1 | 2 | 3 | 4 | 5 |

J F M A M J J A S O N D

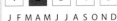

How to find

■ **Timing** The first Bar-tailed Godwits arrive in mid-summer, but numbers peak between November and February, with a gradual return migration through spring. It is best to search for this species around high tide, when views will be closer.

■ **Habitat** Unlike the Black-tailed Godwit, this species favours saltwater environments, from sandy seashores to exposed mudflats in estuary mouths. It may also visit brackish coastland marshes and lagoons but is very scarce inland.

■ **Search tips** Often this species will be the largest brown wader on the beach or estuary except for Curlews, from which it is easily told at a distance by its straight rather than curved bill. Scan the shore and mudflats, and also keep an eye on the sea itself for birds flying past. The state of the tide will determine how close a view you can obtain. If birds are being pushed high up the beach by the incoming water, be especially careful not to disturb them, as they may have to fly considerable distances to find another place to feed, which depletes their precious energy reserves.

Super sites

1. Firth of Forth
2. RSPB Morecambe Bay
3. RSPB Dee Estuary
4. Humber Estuary
5. The Wash
6. RSPB Snettisham
7. RSPB Titchwell
8. RSPB Minsmere

WATCHING TIPS

As with the other waders that specialise in the inter-tidal zone, a good strategy for watching these birds is to pick a spot where they will be pushed towards you by a rising tide, and if you remain still and quiet (and not too close to the high water mark) you should be rewarded with good views. Some reserves have suitably placed hides. Very high tides that cover all the feeding habitat will force the birds to roost elsewhere, and again many of these key roosts are on protected land and overlooked by hides. If you are unsure about godwit identification, concentrate on leg length and face pattern (Bar-tailed has shorter legs and a longer supercilium).

Whimbrel

Numenius phaeopus

Closely related to the Curlew, the Whimbrel is a scarcer bird and only present in the spring and summer. It breeds in the far north-west of Scotland and visits coastlines elsewhere (except north-west England and Scotland) on migration.

Length 37–45cm

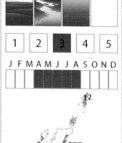

1	2	**3**	4	5

J F M A M J J A S O N D

How to find

■ **Timing** Whimbrels are more readily found on spring than on autumn migration, with several sites regularly hosting flocks on stop-off stays. They are present on their breeding grounds through mid-summer.

■ **Habitat** Migrating birds seek out muddy coastlines, preferring more sheltered saltmarsh, brackish marsh and estuarine creeks to the open seashore. They do also visit inland sites but generally only drop in for no more than a day and sometimes less than an hour, while stays on the coast may be for several days. Breeding birds use rough open grassland with pools, flows and lochans.

■ **Search tips** When at the coast in spring, listen for the distinctive seven-note whistled call, which will give away a bird in flight. Scan muddy shores around estuaries, especially sheltered creeks. Though very similar to the Curlew, it tends not to associate with this species. It is, however, a gregarious bird when on passage and large flocks may be encountered. On the breeding grounds, Whimbrels may be quite visible, one bird standing sentinel while the other is hidden on the nest scrape.

Super sites

1. RSPB Yell
2. RSPB Fetlar
3. RSPB Saltholme
4. Brockholes Wetlands
5. Wheldrake Ings
6. RSPB Rainham Marshes
7. Rye Harbour
8. RSPB Dungeness

WATCHING TIPS

Whimbrels are easily confused with Curlews, especially as the latter shows so much variability in bill size. Look carefully at the bill shape, which is smoothly decurved in the Curlew but looks more 'straight with a kink' in Whimbrel, as well as the face pattern. Regular Whimbrel roosts can be watched in springtime; many are on nature reserves and viewable from hides so there is no risk of disturbance. When walking in Whimbrel country in summer, be aware that these birds will mob and perhaps even strike you if you stray too near the nest. If you see a Whimbrel with a dark rump, it could be the Siberian subspecies, or a North American Hudsonian Whimbrel.

Curlew

The largest wader in Britain, this widespread species is a common breeding bird of northern uplands and may be found in winter anywhere around the coastline, although its breeding population has declined in recent years.

Length 50–57cm
Wingspan 110cm

1	2	3	4	5

J F M A M J J A S O N D

How to find

■ **Timing** Curlews are in evidence on their breeding grounds between April and July. At other times look for them on and near the coast – numbers build up through autumn to peak in mid-winter.

■ **Habitat** In winter most Curlews stick close to the coast, feeding on mudflats, estuaries and the seashore itself, where they can probe the soft ground for worms. They will also feed in wet fields near the coast, especially when chased away from the inter-tidal zone by incoming water. Breeding habitat tends to be open and often elevated ground – heathland, moorland and rough grassland, with wet boggy areas for feeding.

■ **Search tips** In breeding habitat, listen for the beautiful bubbling song, which carries a considerable distance and is often the first clue of a bird's presence. Otherwise, breeding birds can be very discreet – when walking in suitable habitat scan ahead as you go. In winter they are easy to find on and around the seashore, their size making them noticeable from a distance. Be aware that in flight they look at first glance rather like juvenile gulls.

WATCHING TIPS

Curlews are shy and wary birds at all times, so bear this in mind when watching them. If you are walking away from footpaths in breeding habitat, be ready to change course to give birds plenty of space, and watch feeding birds on the shore from a good distance. In late summer when juvenile Curlews are arriving at the coast, beware confusion with Whimbrels, as young male Curlews may have very short bills compared to the longest-billed adult females. Birds on the breeding grounds give a lovely song, but equally evocative onomatopoeic call may be heard all year round.

Spotted Redshank

Tringa erythropus

The taller, more elegant and scarcer cousin of the Common Redshank, this species is a passage migrant and winter visitor to Britain, visiting most coastlines around England, Wales and southern Ireland, although usually in rather small numbers.

Length 29–31cm

| 1 | 2 | **3** | 4 | 5 |

J F M A M J J A S O N D

How to find

■ **Timing** The main autumn passage of Spotted Redshanks through Britain is in late August and September, although failed breeders appear earlier, in the full finery of their breeding plumage. After the passage period, a proportion remain through the winter.

■ **Habitat** These birds mainly visit coastal marshlands, both salt and freshwater, with stretches of shallow water and muddy or vegetated shorelines. They are uncommon inland, and are not inclined to feed on the seashore.

■ **Search tips** Spotted Redshanks are usually seen feeding out in the open water, where they will wade belly-deep and pick morsels of food from the surface. They are usually seen singly, or in very small groups. Their feeding behaviour is rather similar to Black-tailed Godwits, and (in winter plumage) they approach this species in size and general appearance so double-check any godwits you see. They will often share their habitat with Common Redshanks, which allows for side-by-side views that make it easy to see the difference between the species.

Super sites

1. RSPB Blacktoft Sands
2. The Wash
3. RSPB Titchwell
4. RSPB Minsmere
5. Abberton Reservoir
6. RSPB Elmley Marshes
7. Oare Marshes
8. Beaulieu Estuary

WATCHING TIPS

Finding the lone Spotted Redshank among many Common Redshanks is easy, provided the Spotted Redshank is actually present. Look for a longer-legged bird that wades more deeply, has a longer bill with a 'drooping' tip, and more boldly-patterned face. Once you have found your target, watch to see the differences in behaviour. Spotted Redshanks in spring and summer are unmistakable with their beautiful velvety black plumage dusted with whitish speckles, and it is well worth seeking them out if you have only seen them in their pale winter plumage.

Common Redshank

This is one of our most common breeding waders, and in winter its numbers almost double as more arrive from further north and east. It breeds near and winters on coasts all around Britain, but inland is more common in the north.

Length 27–29cm

1	2	3	4	5

J F M A M J J A S O N D

How to find

■ **Timing** Common Redshanks are on the coast year-round, although much more numerous and perhaps easier to see in winter than summer. Inland breeding birds are only on their territories in spring and summer.

■ **Habitat** Low-lying, quiet stretches of coast with wet pasture provide nesting habitat for Common Redshanks, with similar habitat inland being used in northern England, central Ireland and Scotland, more rarely elsewhere. In winter they are more strictly coastal and will feed on the seashore, in salt or freshwater marshland, in estuaries, pools, creeks, on pasture (especially with pools and ditches) and on mudflats.

■ **Search tips** Breeding Common Redshanks often survey their territory from vantage points, flying away with loud calls if you get too close. The call is also a feature of wintering habitat – when you hear it, look out for a medium-sized brown wader with prominent white trailing edges to the wings. Outside the breeding season this species is very gregarious and also readily associates with other waders, but flocks join up quickly in the air if a roost is disturbed.

WATCHING TIPS

Common Redshanks are nicknamed the 'sentinels of the marshes', because of their habit of loudly warning other marshland birds that a human is approaching with a singing '*teu, teu*' call. However, in areas much used by walkers or other visitors they can be quite tolerant and allow you to get quite close. With patience, once you have worked out where a breeding territory is, you could see the parents with small chicks in late May or June, if you check pools, ditches or anywhere with shallow standing water which will attract the flies on which they feed.

Greenshank

Tringa nebularia

Although there are some Greenshanks in Britain at any time of year, they comprise separate populations of breeding birds, passage migrants and overwintering birds. They breed in the Highlands and Hebrides, but are otherwise generally coastal in distribution.

Length 30–32cm

| 1 | 2 | 3 | 4 | 5 |

J F M A M J J A S O N D

How to find

■ **Timing** Greenshanks are on their breeding grounds from April to August. Passage migrants, which mainly breed further north than 'our' birds, are passing through in April to May, and returning mainly between July and September.
■ **Habitat** Greenshanks do most of their feeding in still water in which they can wade. Coastal wetlands, both fresh and salt, will be used, as will slow-moving coastal rivers and estuarine creeks and pools. Passage migrants also make short stop-offs at inland wetlands. The breeding habitat is rough, boggy upland moor.
■ **Search tips** A typical sighting of a wintering or migrating Greenshank is a largish but strikingly slim, rather pale wader working its way through belly-deep water, picking at flies and surface-swimming insects as it goes. In Britain at least, it is not a highly gregarious species, either with its own kind or other waders. The fluty three-note call will draw your attention to a departing Greenshank, and if you watch carefully to see where it settles and then approach slowly and with care you could enjoy good views.

Super sites

1. RSPB Trumland
2. Eden Estuary
3. RSPB Leighton Moss
4. RSPB Saltholme
5. RSPB Blacktoft Sands
6. RSPB Titchwell
7. Stiffkey Fen
8. Cley Marshes
9. Poole Harbour
10. RSPB Cliffe Pools

WATCHING TIPS

Many coastal nature reserves have hides overlooking the kinds of wetlands that attract Greenshanks. Sit quietly and wait – Greenshanks cover a lot of ground when feeding and could well come very close. When walking in the breeding habitat be aware of the risk of disturbance and change course if you flush a bird or see signs of agitation. Early in the breeding season you could see territorial males performing their song-flight, which involves high-flying and a tumbling display. This species is sometimes confused with Green Sandpiper, but is always paler and looks big-headed.

Green Sandpiper

A smallish wader with strongly contrasting dark-and-white plumage, the Green Sandpiper is a passage migrant and winter visitor, mainly to southern parts of England, Wales and Ireland. It also has a tiny breeding population in Scotland.

Length 20–24cm

| 1 | 2 | 3 | 4 | 5 |

J F M A M J J A S O N D

How to find

■ **Timing** Green Sandpipers start to arrive in mid-summer, their arrival coinciding with the shrinkage of ponds and lakes, exposing mud for feeding. Some birds remain through the winter.

■ **Habitat** This is a wader of fresh waters, which may include slow rivers, gravel pits, lakes, reservoirs and even small temporary ponds, often well inland. It prefers sites with some marginal vegetation, offering cover as it works its way along the shoreline.

■ **Search tips** Many sites are used year on year, with the only uncertainty being the date when the first returning migrant will appear. However, the bird can be difficult to spot, as it has rather discreet habits and does not form flocks – there are rarely more than a handful present at any one site. Scan margins and shorelines, looking for the bright white belly and gentle bobbing motion that often gives the bird away. Breeding birds should not be searched for, owing to the tiny size and fragility of the UK breeding population.

WATCHING TIPS

Green Sandpipers are shy, easily alarmed and so best watched very quietly from a hide. You will need patience, but even if the bird is at the far side of the pond when you start watching, it is likely to eventually make its way closer, albeit at a frustratingly slow pace – this is a bird that covers its feeding ground very thoroughly. Passage migrants will often stick around a single smallish body of water for days or weeks, so keep going back and eventually you should manage to obtain point-blank views. If you only get a brief view of a flying bird, the dark upperside with contrasting white rump should be enough to clinch the identification.

Wood Sandpiper

Tringa glareola

This very attractive small but elegant wader, rather similar to the Green Sandpiper, is mainly a passage migrant to southern and eastern England, although it does also maintain a very small breeding population in the Scottish Highlands.

Length 19–21cm

| 1 | 2 | **3** | 4 | 5 |

J F M A M J J A S O N D

How to find

■ **Timing** Migrating Wood Sandpipers start to appear at wetlands from as early as July, with failed breeders first, then juveniles and other adults. Most will have moved on by mid-autumn. Breeding birds are on their territories between April and July, but passage migrants in spring are rarely seen.

■ **Habitat** The Wood Sandpiper has similar habitat requirements to the Green Sandpiper, preferring salt water to fresh, with at least some marginal vegetation. The Wood Sandpiper will, however, use more open and coastal sites than the Green Sandpiper. It breeds on wet, undisturbed grassland.

■ **Search tips** Look for a medium-sized but slight wader, usually on its own, picking its way along shorelines of suitable water bodies. The Wood Sandpiper is often confused with the Green Sandpiper, but as well as being slimmer and longer-legged, has paler and more spangled upper parts and a stronger facial pattern. The breeding population is very small but birds sometimes breed at RSPB Loch Insh in the Highlands, where they may be noticed in territorial display.

Super sites

1. RSPB Insh Marshes
2. Spurn Head
3. RSPB Titchwell
4. RSPB Minsmere
5. Hornsea Mere
6. Abberton Reservoir
7. RSPB West Canvey Marshes
8. Sandwich Bay

WATCHING TIPS

Much of the advice given for Green Sandpiper also applies to this species. Because it is rather scarcer, and may be confused with Green Sandpiper and other species, you should observe carefully and take notes or photos to confirm identification if you are not already familiar with the species. It is a particularly elegantly proportioned and graceful wader, features that help distinguish it from dull-legged juvenile Common Redshanks. Breeding Wood Sandpipers in Scotland may be found away from the known areas, and if so the local bird recorder should be informed.

Common Sandpiper

This distinctive small wader is quite a common breeding bird in northern England, Scotland, North Wales and the north-west of Ireland and Northern Ireland. Elsewhere it occurs as a passage migrant. It also overwinters here in small numbers.

Length 18–20cm

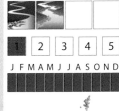

1	2	3	4	5

J F M A M J J A S O N D

How to find

■ **Timing** Migrants return from March, often stopping off at sites in the south before moving on to their breeding grounds, where they will be until July or August. Juveniles and return migrants stop off again through August to October, leaving just a few winterers by mid-autumn.

■ **Habitat** These birds breed alongside stony-shored waters, often quite fast-flowing rivers, or lakes and lochs on high ground with few other breeding water birds. On passage they will visit all kinds of wetlands, from reservoirs to sewage farms, although they often show a preference for spots with water flow, including weirs and dams.

■ **Search tips** A piercing volley of calls gives away a Common Sandpiper as it flies low on bowed wings across a quiet loch. This is a classic way to encounter the species, and with patience you can find the calling bird after it has landed. When looking for the species generally, scan any shorelines that have a rocky or gravelly substrate, and also check the shallows. In early summer look out also for the downy chicks which will be accompanying those adults that have successfully nested.

WATCHING TIPS

This sandpiper is charming to watch as it walks along a loch shore or riverside, picking at the water's edge and ceaselessly bobbing its long body. Its behaviour on passage migration is rather similar – it fly-catches around disturbed water and scuttles along the shingly shores and islands of gravel pits and reservoirs. Identification is usually straightforward – look for the sandy plumage, the white 'spur' on the chest above the wing bend, the rather horizontal stance and the long tail. This is rather a solitary species at all times, unlikely to be found in mixed wader flocks.

Turnstone

Arenaria interpres

The Turnstone, an unusual small wader, is primarily a passage migrant and winter visitor to Britain although a few non-breeders spend summer here. Its distribution is strictly coastal, on almost any coastline except in the far north-west.

Length 22–24cm

| 1 | 2 | 3 | 4 | 5 |

J F M A M J J A S O N D

How to find

■ **Timing** Migrating Turnstones from mainland Europe pass through in July and August, and again on their return journeys in mid-spring. Their presence overlaps with that of the later arriving wintering birds, which originate from Canada and Greenland.

■ **Habitat** Turnstones are beach birds, and will use shingly, rocky and sandy shores with equal enthusiasm. They frequently gather to pick over the shelly sea life that clusters around the bases of piers, harbour arms and groynes, and they will also feed on fields and by coastal lagoons. In towns they may hunt worms in ornamental gardens or even scavenge scraps at people's feet.

■ **Search tips** Turnstones are often very confiding and approachable, and so can be very easy to see, although in winter plumage they blend in surprisingly well with rocks and shingle. Scan along the shorelines – you may find a bird's bright orange legs are the first things to draw your attention. In flight their complex black-and-white pattern is eye-catching. Although summer birds are scarce (especially further south), they are worth seeking out for the attractive breeding plumage.

WATCHING TIPS

Turnstones get their name from their feeding behaviour, and it is interesting to watch a flock working its way along a rocky beach, investigating around and under everything as they go. They will also probe soft ground for worms and other burrowing creatures and are not above snatching a fellow Turnstone's catch right out of its bill. In many seaside towns this behaviour can be watched at point-blank range; the birds will even scavenge for crumbs at tourists' feet. Every Turnstone group should be checked thoroughly for Purple Sandpipers, which have similar ecological needs and often flock with them.

Phalaropus lobatus   **Red-necked Phalarope**

This is an exquisitely dainty small wader that spends most of its time swimming rather than wading. It is a rare breeding bird of the northern Scottish islands, and a similarly rare passage migrant, mainly on the east coast of England.

Length 18–19cm

| 1 | 2 | 3 | **4** | 5 |

J F M A M J J A S O N D

How to find

■ **Timing** To see breeding birds, it is best to visit in May or June. Return migration begins from late summer, with September being the best month to look for a passage migrant. There is no noticeable spring passage.

■ **Habitat** Red-necked Phalaropes breed around small, shallow pools in boggy open countryside. On passage, they are usually found on coastal waters, and sometimes at sea in sheltered bays, however their very small size means that they are easily overlooked on a seawatch.

■ **Search tips** The RSPB's reserve on Fetlar in Shetland offers a wonderful chance to see this very rare bird on its breeding grounds, without the risk of disturbance. The birds are easy to see from the roadside or the hide, and can be watched at leisure going about their business. During the autumn migration period, it is worth checking any coastal lake or marsh for the species, especially following easterly winds – scan all areas of open water, bearing in mind that you are looking for a very small swimming bird. If the weather stays unsettled, migrants may stay several days.

Super sites

1. RSPB Fetlar
2. RSPB Loch na Muilne
3. Welney WWT
4. RSPB Titchwell
5. Stiffkey Fen
6. Cley Marshes WT
7. RSPB Minsmere

WATCHING TIPS

Watching nesting Red-necked Phalaropes reveals several interesting and unusual traits, notably the reversed male/female roles. Females are more colourful than their mates and take the responsibility of establishing and defending the territory, while males take care of the eggs and chicks. Their feeding behaviour is also fascinating – they swim strongly, and spin rapidly on the spot to stir up invertebrate prey which is then snapped up in the needle-fine bill.

Grey Phalarope

Phalaropus fulicarius

The Grey Phalarope neither breeds nor winters here, but southbound migrants, passing mainly to the west of Britain, are sometimes blown towards our shores and even inland. Most sightings are of birds at sea but it occasionally rests on lakes or reservoirs.

Length 21–22cm

| 1 | 2 | 3 | 4 | 5 |

J F M A M J J A S O N D

How to find

■ **Timing** For your best chance of seeing a Grey Phalarope, keep an eye on weather forecasts through the autumn. Westerly gales will almost always drive a few birds inland, with more migrating past just offshore. While the west coast is usually best, there may be sightings along the east coast too, especially if the gales have a northerly component.

■ **Habitat** Grey Phalaropes spend their winter at sea and can tolerate quite rough conditions. Onshore winds will produce sightings closer to land, or even upriver. When pushed well inland they may spend hours or days recovering on reasonably-sized water bodies of various kinds.

■ **Search tips** Seeing this species on a seawatch is quite a challenge, as it is a small bird and its grey-and-white plumage does not stand out clearly against the waves. The usual seawatching advice applies – pick an elevated spot where you can see between the wave crests but not so high that you will be too distant to pick up a small bird flying by. After westerly gales, check your local lake, gravel pit or reservoir, looking for a small but stocky, mainly white bird, swimming buoyantly.

WATCHING TIPS

Finding an inland Grey Phalarope gives you the chance to watch this rare bird in action, sometimes at close range as, like other phalaropes, it may be very confiding. Like the Red-necked Phalarope, it feeds by spinning on the water to disturb and then snap up its prey. While seawatching may not produce sightings of this quality, on occasion the quantities involved can be considerable, with exceptional days producing counts of several hundreds from west-coast headlands.

Pomarine Skua

This skua breeds in the Arctic and winters off west Africa, a migratory path that may take it past our coastlines. Most spring sightings are in the south or far north, and in autumn they are seen down the east coast.

Length 46–51cm
Wingspan 125–138cm

| 1 | 2 | 3 | 4 | 5 |

J F M A M J J A S O N D

How to find

■ **Timing** May is best for spring sightings, when birds in magnificent adult plumage fly past south coast headlands and skim Shetland and the Western Isles. In autumn, headlands down the east coast produce sightings of mainly juveniles. An onshore wind will increase the chances of sightings in all cases – on very still days birds are likely to go by further out to sea.

■ **Habitat** Pomarine Skuas are usually seen flying by at sea. They may be attracted by other seabirds, which they chase to steal their food. Exhausted or storm-driven birds are sometimes found resting on beaches or fields just inland – very occasionally they may take refuge on inland lakes or reservoirs.

■ **Search tips** The usual seawatching advice applies – pick a day when the wind is right, choose a somewhat elevated spot on a headland, and wait. Often there are more sightings in the early morning than later on – good days will see small groups. Look out for dark, long-winged but powerfully built, barrel-chested birds with quite agile and relaxed flight.

Super sites

1. Esha Ness
2. Aird an Runair
3. Bowness-on-Solway
4. Flamborough Head
5. RSPB Skua and Shearwater Cruise from Bridlington
6. Porthgwarra
7. Portland Bill
8. Seaford Head
9. RSPB Dungeness

WATCHING TIPS

In autumn especially, identification of this species can be problematic, as juvenile Pomarine Skuas are hard to separate from juvenile Arctic Skuas. Concentrate on the general shape, plumage tones and the size and shape of the white wing flash. With good clear views, the bill pattern can also be helpful for identification. Spring adults present no such problems, with their unique spoon-shaped central tail feathers. Pelagic trips in spring and autumn offer a chance of very close views. Seeing spring Pomarine Skuas is almost a competitive sport among dedicated sea-watchers, with the most patient reaping the rewards.

Arctic Skua

Stercorarius parasiticus

This graceful seabird, our most frequently seen skua, breeds in Orkney and Shetland and the far north mainland of Scotland, and migrates along most coastlines around the British Isles. Both adults and juveniles occur in pale and dark morphs.

Length 37–44cm
Wingspan 110–125cm

1	2	**3**	4	5

J F **M A M J J A S O** N D

How to find

■ **Timing** To see Arctic Skuas on their breeding grounds, visit between May and July. Birds may be seen elsewhere around the coast during the spring and autumn migration periods, and occasionally in winter.

■ **Habitat** This skua nests on the ground on open moorland in loose colonies. In the breeding season adults will commute to nearby seabird colonies, especially tern colonies, obtaining much of their food by pursuing the terns to force them to drop their catch. Outside the breeding season, it is seen at sea, with onshore winds bringing it closer inshore. It may visit coastal lagoons where terns are feeding, and occasionally is driven inland by storms.

■ **Search tips** Arctic Skua colonies are quite obvious when you are in the right areas. Rather than relying on concealment as protection against predators like some ground-nesters, these skuas consider attack the best form of defence and rise up to try to see off an intruder. At tern colonies, watch out for an agile dark bird intercepting the terns at sea – and look out for a similar sight in autumn when there are migrating terns making their way along the coast.

Super sites

1. RSPB Fetlar
2. Fair Isle
3. RSPB North Hill
4. RSPB Birsay Moors
5. Carnsore Point
6. RSPB Titchwell
7. St Ives Island
8. Pendeen
9. Porthgwarra
10. Portland Bill
11. RSPB Dungeness

WATCHING TIPS

Visitors to an Arctic Skua colony are likely to be subjected to noisy dive-bombing attacks, with several birds joining in the mobbing. Although the birds seldom make physical contact, you should certainly give them a wide berth, and wearing a hat is a good idea! When you're seawatching in autumn, juvenile skuas present real identification challenges. Arctic is the commonest species, but may be confused with both Pomarine and Long-tailed Skuas. Record as much detail as you can, but accept that some cannot be safely identified. Dark-morph adults are at least always indistinguishable from adult Long-tailed Skua, which has no dark morph.

Stercorarius longicaudus

Long-tailed Skua

One of the most beautiful seabirds in the world, the Long-tailed Skua breeds in the Arctic and is a rare passage migrant past the British coastline, with most records coming from the east coast and in the autumn passage period.

Length 35–41cm
Wingspan 105–117cm

| 1 | 2 | 3 | 4 | 5 |

J F M A M J J A S O N D

How to find

■ **Timing** The likeliest period to see Long-tailed Skuas from east coast headlands is between late August and mid-September. Spring birds are more often seen around the north Scottish coast and islands. Going to look around high tide and when onshore winds are blowing will improve the chances of closer views.
■ **Habitat** Most sightings are of birds going by offshore, but occasionally tired or storm-blown birds will come ashore to rest on beaches or in coastal fields, and may remain for a few days.
■ **Search tips** Even with the best-laid plans, you need a good dose of luck to see this species, particularly to see it at its best in adult plumage. Most autumn sightings are of tricky-to-identify juveniles, often some distance out at sea. Patience, frequent visits to good sites under suitable weather conditions, and having a good telescope (with which you are fully familiar and confident) will all help. Arrive at the headland early, and if possible join forces with other birdwatchers so there are more eyes scanning the water. Any small-looking, very agile skua should be scrutinised with particular care.

Super sites

1. Flamborough Head
2. Sheringham
3. RSPB Dungeness

WATCHING TIPS

The adult Long-tailed Skua is unmistakable, and while sightings in spring are fewer than autumn, if you watch in spring there is a better chance that any you see will be adults. In the far northern islands returning adults may even make landfall. In autumn, the identification of this skua is usually a challenge – a likely candidate is a bird that looks especially small and long-tailed, with only a small white wing flash and rather cold plumage tones. Unfortunately, not all skuas can be identified, especially when viewing conditions are poor. Try pelagic trips to improve your chances of closer views.

Great Skua

Catharacta skua

The largest skua, this heavy-set and powerful bird has no tail streamers and looks rather similar to a large juvenile gull. It breeds in north Scotland – on the mainland and islands – and migrates along our coastlines in spring and autumn.

Length 50–58cm
Wingspan 132–140cm

1	2	3	4	5

J F M A M J J A S O N D

How to find

■ **Timing** Visit the breeding grounds between May and July to see the adults, which will also visit any nearby cliff or beachside seabird colonies during the breeding season. Migrating birds are best looked for from headlands, especially down the east coast in autumn.

■ **Habitat** These birds nest on open moorland, either grassy or heathy, in loose colonies, picking spots that give them a good view of the surrounding area. They are common visitors to cliffside seabird colonies, where they harass gulls and Gannets, forcing them to disgorge their food. They also hunt smaller seabirds such as Puffins. Migrants travel offshore, the line they take depending on weather conditions.

■ **Search tips** As with Arctic Skuas, these birds make no real effort to conceal themselves when nesting, and instead take to the air to drive off potential threats. Adults will often be wheeling around in the vicinity of the colony. When watching breeding seabirds from clifftops, watch out for Great Skuas chasing Gannets and Kittiwakes as they return from their fishing trips. The large white wing-flash stands out from a distance.

Super sites

1. RSPB North Hill
2. RSPB Hoy
3. Redpoint
4. RSPB Troup Head
5. Flamborough Head
6. Spurn Head
7. Pendeen
8. Porthgwarra
9. RSPB Dungeness

WATCHING TIPS

Great Skuas are vigorous in their defence of their nests, and you should avoid going too close for your own safety as well as the birds' peace of mind. Wear a hat, as they will aim to kick or claw at your head. To watch the skuas in piratical mode, spend time at a north coast cliffside colony and you could witness some dramatic chases, as the skua relentlessly pursues its chosen victim out over the sea. With luck, similar behaviour could be observed from a seawatch or pelagic trip. Sometimes migrating Great Skuas may spend a day or two on the beach, and can be approachable.

Larus melanocephalus

Mediterranean Gull

This beautiful gull began to colonise south-eastern England in the second half of the 20th century, associating with colonies of Black-headed Gulls. It has spread slowly north and west along the coast, and ranges more widely in winter.

Length 36–38cm
Wingspan 92–100cm

1	2	**3**	4	5

J F M A M J J A S O N D
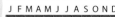

How to find

■ **Timing** Mediterranean Gulls may spend winter near their breeding colonies, but breeding activity proper begins around March. By mid-September the colonies are quieter, and birds are dispersing, with some moving along the coast and others visiting sites inland.

■ **Habitat** Most breeding pairs at present nest among colonies of Black-headed Gulls. Colonies are typically at the coast, on thinly vegetated shorelines and islands of lakes and lagoons. At other times Mediterranean Gulls spend time on beaches, resting on calm seas, at rubbish tips and at lakes and reservoirs, usually associating with other gulls. They may visit town parks and busy seaside town promenades and beaches.

■ **Search tips** The coal-black head, large red bill and prominent white eye-ring helps the adult Mediterranean Gull stand out among the more numerous Black-headed Gulls in breeding colonies. Birds in flight often give themselves away with a distinctive call. Among flying flocks or winter birds resting on the sea, look out for the Mediterranean Gull's white wing-tips. Odd Mediterranean Gulls are often found within large gatherings of gulls, so check your local gull roost regularly.

Super sites

1. Great Yarmouth
2. RSPB Minsmere
3. Southend-on-Sea
4. Plymouth
5. RSPB Radipole Lake
6. Rye Harbour Nature Reserve
7. RSPB Dungeness

WATCHING TIPS

Life at a gull colony is frenetic and noisy. From a hide you can watch the comings and goings at leisure, and compare Black-headed and Mediterranean Gulls side by side. Certain coastal towns in winter attract many Mediterranean Gulls, and this gives you the chance to compare adult and subadult plumages at close range, and alongside other species (first-winter Mediterranean Gulls are more similar to the larger species than to Black-headed). Recording the species' spread is important, so look out for signs of breeding behaviour.

Little Gull

Hydrocoloeus minutus

A tiny, dainty and rather tern-like gull, the Little Gull is a passage migrant and winter visitor to Britain, visiting coastlines around England, Wales, south-east Scotland and southern Ireland. Certain sites, especially in north England, attract large concentrations.

Length 25–27cm
Wingspan 75–80cm

| 1 | 2 | 3 | **4** | 5 |

J F M A M J J A S O N D

How to find

■ **Timing** This species is most numerous offshore around Britain between mid-summer and early spring, though there are sightings all year round. Flocks form at Bridlington Bay, Hornsea Mere and Fife Ness in autumn, and the Crosby area in Merseyside in spring.

■ **Habitat** Little Gulls are often seen flying offshore, but may also come inland, particularly in spring and summer. They will visit coastal lakes and catch emerging flies over the open water. They tend not to join flocks of larger gulls very often, especially inland, but may gather at offshore outflows.

■ **Search tips** In the traditional flocking areas, Little Gulls should be easy enough to find. Autumn birds flock in rafts on the sea, while spring birds hawk gracefully over the lakes in a manner reminiscent of Black Terns. Finding Little Gulls on a seawatch is more a matter of luck, but any area with a strong passage of terns is quite likely to come up trumps. From a distance, the sooty-black underwings are perhaps the best identification feature, along with the small size, round-looking wing-tips and tern-like flight.

Super sites

1. Fife Ness
2. Crosby Marine Park
3. Bridlington Bay
4. RSPB Titchwell
5. Hornsea Mere
6. RSPB Dungeness

WATCHING TIPS

To enjoy more than just a 'fly-past' view of a Little Gull, visit one of the key sites at the right time of year and you should be able to watch them at length. Numbers in the north-west sites vary from year to year, but with luck there will be birds at Crosby Marine Park from mid-April to early May, exploiting the hatch of midges. The Bridlington Bay flock can almost reach four figures, and will include both adults and young birds, giving the chance to study the different plumages of this attractive gull. Juveniles are particularly striking with their complex and bold black-and-white pattern.

Sabine's Gull

This gull breeds in the Arctic and migrates to south of the equator for winter. It may be seen at sea from any coast, although there are more records from the western side of the British Isles than the east.

Length 27–32cm
Wingspan 90–100cm

1	2	3	4	5

J F M A M J J A S O N D

How to find

■ **Timing** Although Sabine's Gull may pass our shores in both migration periods, the autumn period produces far more sightings than spring, and September is the best month. Strong onshore winds push the birds closer to land, and some birds are found inland immediately after stormy periods.

■ **Habitat** This is very much a seafaring gull, mostly seen from coastal headlands. If the winds are not right it will go by well offshore. They may linger with other gulls at warm water outflows and occasionally on the beach. Storm-driven birds head up major rivers, or shelter on large reservoirs or lakes.

■ **Search tips** When there are strong westerly winds in autumn, time spent seawatching from headlands that project into the Atlantic should produce sightings of Sabine's Gull as well as other interesting seabirds. Easterly winds will also push Sabine's Gulls close to east coast headlands. The usual seawatching advice applies – watch from an elevated position, and for this species beware confusion with juvenile Kittiwakes, which look similar from a distance. The day after very powerful autumn storms will almost always turn up a few at inland waters – check larger reservoirs.

Super sites

1. Annagh Head
2. Bridges of Ross
3. Flamborough Head
4. Strumble Head
5. Pendeen
6. St Ives

WATCHING TIPS

Sabine's Gulls are attractive birds and, in the right conditions, can be viewed at close quarters. Storm-driven birds are often fearless and may swim or fly very close to observers – they may remain on their chosen reservoir for a few days waiting for weather conditions to improve before moving on. Pelagic boat trips offer the chance of good views of migrants at sea – because these trips will go well offshore the chance of sightings is (thankfully) not dependent on stormy weather. Even with poor views, the unique 'three-triangle' wing pattern is usually easy to make out.

Black-headed Gull   *Chroicocephalus ridibundus*

The commonest small gull in Britain, the Black-headed Gull can be found inland and on the coast almost anywhere in Britain, often in large numbers. We have a substantial breeding population, but numbers in winter increase seven-fold.

Length 34–37cm
Wingspan 100–110cm

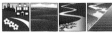

1	2	3	4	5

J F M A M J J A S O N D

How to find

■ **Timing** Black-headed Gulls are in their busy breeding colonies from March until mid-summer. During this period, numbers elsewhere may be low, although there are usually a few subadult and non-breeding birds at the same sites that will host hundreds or thousands in winter.

■ **Habitat** Nesting colonies form both inland and on the coast, on the ground by still or slow-flowing water, such as rivers, lakes, reservoirs or marshes. Lakes with islands are especially popular, offering more protection from ground predators. Colonies may be extremely dense, with pairs nesting within touching distance of each other. In winter, usually found near water, on the coast or inland, but also visits arable fields, and will come for food to town gardens.

■ **Search tips** Finding this gull is rarely a challenge. Colonies are very vocal, relying on mass mobbing rather than concealment to deal with predators. In winter, most town lakes will hold several Black-headed Gulls, which will readily come to food thrown for the park ducks. Gulls seen on farmland will usually include some Black-headed Gulls, and they also join flocks roosting on reservoirs or pillaging rubbish tips.

WATCHING TIPS

A visit to a colony of nesting Black-headed Gulls is always interesting, whether you visit early in the season when pairs are nest-building and bickering over territory, or later on when the parents are feeding their growing chicks. Watching winter gulls in the park is just as engrossing – the gulls will exhibit tremendous aerial skill as they try to catch scraps of bread in mid-air, and at other times they may amuse themselves by playing with fallen leaves or vigorously bathing in the water. Individuals show considerable variation in their moult timings, with some already in 'summer' plumage by December.

Herring Gull

This gull is probably the most familiar British species, as it is the one most likely to nest in an urban setting. It breeds around almost the entire British coastline, and in winter thousands more visit inland sites.

Length 55–64cm
Wingspan 138–150cm

1	2	3	4	5

J F M A M J J A S O N D

How to find

■ **Timing** Many breeding adults remain around their coastal nesting sites all year, with non-breeding birds also lingering through summer if there are good food supplies. Further inland, large feeding flocks congregate at rubbish tips in winter, and roost on reservoirs (in company with several other gull species).

■ **Habitat** This gull nests in elevated positions by the coast, which includes on sea cliffs and also on rooftops in seaside towns. It is a generalist feeder, and may be seen scavenging fish and chip remains on seaside promenades, picking through tideline debris on the beach, and following the plough on arable fields, as well as flocking to rubbish tips.

■ **Search tips** This is an easy bird to find, especially in seaside towns where it is very visible (and audible) and may actually get a bit too close and personal, snatching food from visitors' hands. Rooftop nests are usually quite obvious piles of sticks and straw, and the young birds' constant whistling begging calls are a feature of mid-summer. In winter, flocks tend to have a regular daily routine, feeding in the middle of the day, and returning to a roost site on water in the late afternoon, where they bathe and rest.

WATCHING TIPS

These bold, gregarious and noisy birds have made their share of enemies by living so closely alongside people in seaside towns, but they are unfailingly entertaining to watch and anyone who lives in a seaside town has the opportunity to make a detailed study of their complex behaviour. Two subspecies occur in Britain, and this along with the variability of young birds (full adult plumage takes four years to attain) makes learning Herring Gull plumages a real challenge, but very worthwhile as it is the main confusion species for a number of rarities.

Common Gull

Larus canus

Despite its name, this is by no means our commonest gull, although it does breed in good numbers in Scotland, northern England, Northern Ireland and west Ireland. In winter numbers increase substantially, and its range expands to the rest of Britain.

Length 40–42cm
Wingspan 110–130cm

| 1 | 2 | 3 | 4 | 5 |

J F M A M J J A S O N D

How to find

■ **Timing** Common Gulls will be nesting between April and mid-summer. At the end of summer, numbers away from the nesting grounds start to build up, peaking in mid-winter.

■ **Habitat** The Common Gull nests on the ground, with most nests close to fresh water (including small streams or pools) but it has no strict tie to this kind of habitat. Nests may be single or in colonies. Nest-sites are often well away from human habitation. Many nest along rivers that cut through high ground, or around the shores of remote lakes and lochs. Once young birds are independent, they and the adults form flocks, feeding in fields, on rubbish tips and at the coast. It will flock with other gulls outside the breeding season.

■ **Search tips** Many winter gull flocks, both at the coast and inland, will include some Common Gulls, which stand out as being intermediate in size between Black-headed Gulls and the large species like Herring and Lesser Black-backed Gulls. White-headed, grey-backed gulls seen inland in summer are likely to be this species.

WATCHING TIPS

Although usually outnumbered by other species, the Common Gull will visit town park lakes and at such sites will become quite confiding and approachable, giving you the opportunity to study it close up, as well as compare it alongside other gull species. Common Gull nests are sometimes in areas vulnerable to disturbance, so be careful not to get too close. Common Gull is the main confusion species for the rare Ring-billed Gull, which may turn up in towns and other busy areas, so in winter it is worth checking all Common Gulls carefully for this transatlantic vagrant.

Larus fuscus

Lesser Black-backed Gull

This gull nests mainly on or near the coast around the whole of Britain, and in winter it is also found inland, especially in England and Ireland. It is easily confused with several other species, especially in immature plumages.

Length 52–64cm
Wingspan 135–150cm

1	2	3	4	5

J F M A M J J A S O N D

How to find

■ **Timing** From mid-winter onwards, adult breeding birds will be on their nesting grounds. Throughout winter, Lesser Black-backed Gulls are common inland, in fact they are frequently the dominant species of large gull roosts. Most gulls will go to their roosts in the last two hours before dusk, and depart within two hours of first light.

■ **Habitat** Lesser Black-backed Gulls may nest on cliffs and rooftops (though they are usually outnumbered by Herring Gulls at sites like this) or on open ground. They are becoming more numerous as breeding birds in urban environments. Their breeding range extends further inland than that of Herring Gulls. In winter they will visit rubbish tips, fields, rivers, lakes (including in urban settings) and many remain at the seaside.

■ **Search tips** This gull is usually easy to find, especially in large gull gatherings inland in winter, though on the coast it may be scarcer than both Herring and Great Black-backed Gulls. It may draw attention even from a distance among Herring Gulls because of its more attenuated and elegant outline.

WATCHING TIPS

This gull can pose significant identification problems, as it can be confused with almost all other large gull species likely to be seen in Britain, both common and rare. Therefore, careful study to become familiar with the species in all of its plumages does improve your chances of finding something 'different' among a flock containing many Lesser Black-backed Gulls. Additionally, three different subspecies of Lesser Black-backed Gull occur in Britain (though one is very rare), adding to an already complex situation. Pay particular attention to proportions and the wingtip pattern when faced with a confusing adult gull. Leg and wing colour are also helpful, but sometimes difficult to assess.

Yellow-legged Gull

Larus michahellis

The Yellow-legged Gull was formerly considered a subspecies of the Herring Gull, and replaces it in the west Mediterranean region. It has now been found to be a separate species, and is increasingly found in Britain.

Length 56–67cm
Wingspan 140–158cm

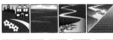

1	2	3	4	5

J F M A M J J A S O N D

How to find

■ **Timing** Yellow-legged Gulls are present in Britain all year round, but the largest numbers are here between late summer and mid-autumn.

■ **Habitat** This is a very rare breeding bird, with the few breeders favouring rooftop nest-sites. Non-breeding Yellow-legged Gulls are attracted to the same sorts of places as other large gulls – rubbish tips, playing fields, farmland fields, reservoirs, large river foreshores and coastal sites.

■ **Search tips** Finding Yellow-legged Gulls can be hard work – many gull flocks through the winter will contain a few but sorting through the commoner species takes time and care. Herring and Lesser Black-backed Gulls are both very similar to Yellow-legged Gulls, especially in subadult plumages. There are some sites where Yellow-legged Gulls form sizeable flocks towards the end of summer, and visiting one of these is an easier way to catch up with the species. It is also worth checking nesting pairs of Herring and Lesser Black-backed Gulls on rooftops, especially on the south-west coast, as Yellow-legged Gulls sometimes form mixed pairs with either of these species.

Super sites

1. Paxton Pits NR
2. Stewartby Lake
3. Chew Valley Lake
4. RSPB Rainham Marshes
5. Outer reaches of River Thames
6. Poole Harbour
7. Pagham Harbour
8. RSPB Dungeness

WATCHING TIPS

It is only since the end of the 20th century that birdwatchers have started to become fully familiar with the identification of this gull in all its plumages. If you are well versed with both Herring and Lesser Black-backed Gulls, your chances of picking out a Yellow-legged among them are much improved. Recording breeding attempts by the species is very important – although it has bred here since 1995, its numbers have barely increased, but there is the possibility that it is overlooked among the commoner species. When holidaying in the Mediterranean, take the opportunity to study this species where it is common.

Larus glaucoides — **Iceland Gull**

A few individuals of this Arctic species wander south to Britain in winter, with most records from northern and western coasts.

L 52–60cm **W** 140–150cm

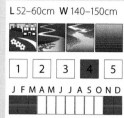

| 1 | 2 | 3 | **4** | 5 |

J F M A M J J A S O N D

How to find it

- **Timing** Mid-winter is the best time to find this gull, although individuals will often stick to the same site for months.
- **Habitat** Iceland Gulls are usually found on the coast in areas such as fishing towns or ports, but may also join roosts further inland.
- **Search tips** This gull is eye-catching for its paleness (in all plumages) and small size. Check through gull flocks on the coast in winter.
- **Super sites 1.** Ullapool Harbour, **2.** Killybegs, **3.** Nimmo's Pier, **4.** North Shields, **5.** Penzance area.

WATCHING TIPS

Iceland Gulls in all plumages are attractive-looking birds and well worth seeking out. The main confusion species is Glaucous Gull – compare size to other nearby gulls. Some of those that visit Britain are 'Kumlien's Gulls', a slightly darker subspecies (possibly a separate species).

Larus hyperboreus — **Glaucous Gull**

This Arctic species looks like a larger Iceland Gull, and has a similar distribution pattern, although it occurs in larger numbers.

L 63–68cm **W** 150–165cm

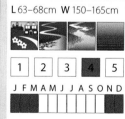

| 1 | 2 | 3 | **4** | 5 |

J F M A M J J A S O N D

How to find it

- **Timing** Most Glaucous Gulls arrive in late autumn, and may spend several months at the same site, only moving on in early spring. Occasionally subadults will oversummer here.
- **Habitat** Most birds settle at coastal sites with a ready food supply and a population of other large gulls with which they flock.
- **Search tips** This large, pale gull is usually noticeable when you scan a gull flock, though beware occasional Herring Gulls with leucism or (in autumn) very worn plumage.
- **Super sites 1.** Thurso Harbour, **2.** Ullapool, **3.** Killybegs, **4.** Nimmo's Pier, **5.** Scarborough Harbour, **6.** Ogston Reservoir, **7.** Penzance area, **8.** RSPB Dungeness.

WATCHING TIPS

Because Glaucous Gulls are usually long-stayers, if one visits a site local to you there will be the opportunity to make repeated visits and watch at length. On the coast, many sites good for Glaucous Gull will also attract Iceland Gulls, allowing for side-by-side comparison of these closely related species.

Great Black-backed Gull *Larus marinus*

The largest of the gulls, this species is fairly common around the coastline of Britain. Numbers increase in winter, when it also visits inland sites, although it remains generally more coastal than the other common large gulls.

Length 61–74cm
Wingspan 150–165cm

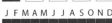

| 1 | 2 | 3 | 4 | 5 |

J F M A M J J A S O N D

How to find

■ **Timing** Around most parts of the coast, Great Black-backed Gulls may be seen year-round, although they are usually outnumbered by Herring Gulls. In winter they join gull flocks using reservoirs and rubbish dumps, though again in relatively low numbers compared to Herring and Lesser Black-backed Gulls.
■ **Habitat** Most pairs nest on wide cliff ledges or among coastal rocks, with smaller numbers on the ground in open, undisturbed countryside (such as moorland). They feed at the usual gull haunts – on beaches, around fishing boats, on rubbish tips – and join communal gull roosts on open fresh water. They will also visit seabird colonies where they prey on young (and small) birds.
■ **Search tips** This gull is easiest to see on the coast. Single birds are often seen on seawatches, strongly flying offshore or loafing on the beach. In all settings its huge size and very dark back and wings are striking among other gulls. Subadults are a little more difficult to pick out plumage-wise but still look noticeably large and heavy-billed. Most large flocks and roosts will hold a few Great Black-backed Gulls – they are usually outnumbered by the smaller species but their size makes them easy to spot

WATCHING TIPS

This is a powerful and predatory bird, more inclined to take substantial large prey than the other species. Anyone watching seabird colonies for any length of time may witness it in action. In seaside towns, Great Black-backed Gulls may be quite approachable, and permit close views, though they are rarely as confiding as the typical seaside Herring Gull. This species can be confused with Lesser Black-backed Gull, especially in winter when darker-backed Scandinavian Lesser Black-backed Gulls visit Britain – study its size and proportions to make a confident identification.

Larus cachinnans

Caspian Gull

Many older field guides do not list this species at all, because it was considered an eastern subspecies of Herring Gull, and because its presence in Britain was not accepted until 2003. Small numbers are now known to occur regularly.

Length 59–67cm
Wingspan 140–158cm

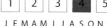

| 1 | 2 | 3 | **4** | 5 |

J F M A M J J A S O N D

How to find

■ **Timing** Most Caspian Gulls in Britain are found in the winter months, but subadult birds in particular may be seen at any time of year.
■ **Habitat** This gull does not breed in Britain but joins flocks of other large gulls at typical sites – rubbish dumps, reservoir or lake roosts and river foreshores have proved the most successful searching grounds.
■ **Search tips** Finding a Caspian Gull will usually require careful searching through a large mixed gull flock, and therefore requires patience and a good level of familiarity with the commoner species. The features that separate it from Herring and Yellow-legged Gulls are all rather subtle but taken together can produce a distinctive-looking bird, although there is much individual variation. Taking photos of any potential Caspian Gull can help to pin down features that are difficult to discern in 'real time'. It may be necessary to make several visits before 'connecting' but birds tend to be quite site-loyal. Bear in mind if visiting rubbish dumps that on days when the site is closed there may be far fewer gulls present.

Super sites

1. Eyebrook Reservoir
2. Blackborough End Tip
3. Stewartby Lake
4. Dix Pit
5. RSPB Minsmere
6. Blyth Estuary
7. RSPB Rainham Marshes
8. RSPB Dungeness

WATCHING TIPS

As a relative newcomer to the British list, the Caspian Gull, a bird native to Central Asia but expanding its range westwards, is still a somewhat unknown quantity. Recording any sightings is therefore especially important, to build up an accurate picture of the bird's distribution in Britain and any trends in its population. Photographs are also proving very valuable for furthering study of its various plumages and moult pattern. A 'classic' Caspian Gull is an attractive bird, well worth the trouble it may take to find one.

Kittiwake

Rissa tridactyla

A unique and beautiful gull, the Kittiwake is our only gull species to nest in busy cliff-face colonies. It breeds at most places around the coast where suitable cliffs exist, and at other times is mainly seen at sea.

Length 39–40cm
Wingspan 95–120cm

1	2	3	4	5

J F M A M J J A S O N D

How to find

■ **Timing** Kittiwakes occupy their breeding cliffs from February until August. Then through the autumn they – and migrants from further north and west – go by our shores in large numbers. The winter months are quieter, though Kittiwakes may still be seen offshore from most points along the coast.

■ **Habitat** Nesting cliffs are usually precipitous and composed of hard stone such as limestone. Kittiwakes can use quite narrow ledges and pack their nests quite close together, often alongside other cliff-nesters like Guillemots. There are also a few colonies on buildings. They visit low rocks and other objects projecting from the sea to collect seaweed (the main nesting material). Passage birds are usually seen offshore, rarely coming to beaches, and are very uncommon inland.

■ **Search tips** At their colonies, Kittiwakes are obvious, wheeling around the cliffs and giving incessant loud calls. They can usually be seen from the shore within a few miles of a colony. In autumn, seawatching from any headland should produce some sightings – also check gatherings of gulls and terns offshore around warm-water outflows.

Super sites

1. RSPB Troup Head
2. RSPB Fowlsheugh
3. Farne Islands NT
4. RSPB Mull of Galloway
5. RSPB Bempton Cliffs
6. RSPB Ramsey Island
7. Seaford Head

WATCHING TIPS

Visiting a seabird cliff at the height of activity is an essential birdwatching experience, and Kittiwakes are a key component of the spectacle. They are wonderfully graceful in flight, and extraordinarily noisy. Regular visits to a colony through spring and summer will allow you to observe the whole breeding cycle, from competition for sites to nest-building, and the growth of the chicks from tiny white bundles of fluff to beautifully marked juveniles embarking on their first flights. Pelagic trips in late summer and autumn can provide wonderful close views of both adult and young Kittiwakes.

Sternula albifrons

Little Tern

A small, short-tailed tern with a rather jerky flight, this species breeds in scattered colonies on beaches around parts of the British coastline, especially south-east England, west Scotland, north Wales, and only sparsely in Northern Ireland and Ireland.

Length 22–24cm
Wingspan 48–55cm

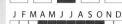

| 1 | 2 | 3 | 4 | 5 |

J F M A M J J A S O N D

How to find

■ **Timing** Little Terns arrive back on their breeding colonies in April. By the end of August adults and juveniles will have moved off and may be seen migrating offshore through early autumn.

■ **Habitat** This species nests mainly on shingle beaches. Colonies may favour one area one year but abandon it the next and occupy a new site several miles away. They are very vulnerable to disturbance and predation, and many colonies are protected by conservation bodies. They fish mainly over the sea adjacent to beaches, quite close inshore where the water is shallow.

■ **Search tips** Anywhere on the coast with shallow and sheltered water could attract feeding Little Terns, although during the breeding season the nearer you are to a colony the better your chances of a sighting. Little Terns have an erratic flight which is noticeable among the more graceful larger terns – they also have a rather top-heavy look, and a distinctive rasping call. Some colonies can be visited, making it easy to see and watch the birds.

Super sites

1. Sands of Forvie
2. Kilcoole
3. RSPB Hodbarrow
4. Blakeney Point
5. RSPB Minsmere
6. Langstone Harbour
7. Rye Harbour

WATCHING TIPS

Exercise care when watching Little Terns at their colonies, to avoid disturbance. Life on the colony is furiously active. Early in the season the males entice females with fish into an aerial chase. Small chicks are often preyed upon by Kestrels so will take shelter when they can under rocks, driftwood or other beach debris. Some colonies are protected by electrified fencing to deter ground predators. At some sites, accurate decoy model Little Terns are used to try to attract the birds – these can be confused with the real thing at a glance!

SEABIRD COLONIES

With its tremendous length of coastline, the British Isles has proportionately more breeding seabirds than other European countries, and is of international importance for several species. Cliff faces, coastal moors, beaches and even seaside buildings all provide homes for seabirds of one kind or another, and a trip to one of these great 'seabird cities' is a real highlight of birdwatching in Britain.

Cliffside communities

A sheer granite cliff face may be desolate in winter, with only a few Jackdaws and the occasional Peregrine Falcon wheeling around the bare rocks. Come the end of winter, though, and the cliff-nesting birds start to arrive. Within a few weeks the cliff is transformed, with every ledge, crack, cranny and crevice home to noisily calling seabirds. Making sense of the mass of bird life may take you a little while, so find a comfy spot with a good view and start scanning.

There may be few birds on the grassy tops, but among them could be everyone's favourite, the Puffin, nesting in burrows dug into the crumbly soil. They will also be on plateaux further down the cliff if there are any large enough. Some plateaux may also be carpeted with Gannets, the largest cliff-nesters, and Fulmars tuck themselves into crevices high up on the slopes.

The sheer faces will be patterned black-and-white with Guillemots, which can nest on the narrowest ledges and have little need for much personal space. Where the ledges widen, you'll find Razorbills, fewer and further between and using more spacious sites than their close relatives, and also Kittiwakes, beautiful gulls whose constant screaming calls provide the bulk of the seabird soundtrack. Further down, on tucked-away ledges close to the sea, there will be Shags, slim black birds with green-glossed wings and hectic curly crests. In some areas, similar spots will be home to Black Guillemots.

Beaches, banks and islands

Another group of seabirds, the terns, habitually nest at sea level, either on the beach itself or on permanent sandbanks, or islands within coastal lakes and lagoons. In some areas they share these nesting sites with some species of gulls, creating a mixed community. Together, they make a formidable force when it comes to seeing off predators – essential given the nature of their breeding sites.

The terns that nest in seaside colonies are Common, Arctic, Sandwich and Little, with the very rare Roseate Tern found at only a handful of sites. Alongside them are often large numbers of nesting Black-headed Gulls, and on the south coast there may also be Mediterranean Gulls. Their colonies are full of activity, excitement and noise from mid-spring to mid-summer, but all falls quiet when the young birds fledge.

The night watch

Two very special seabirds are notable for visiting their breeding colonies at night rather than by day. Both are small and very clumsy on land, easy prey for the larger and more predatory seabirds like gulls and skuas. By confining their activity to the hours of darkness, they avoid this danger. Most colonies are on mammal-free islands – rats and cats can do massive damage to these hole-nesting birds and their chicks. To see Storm Petrels and Manx Shearwaters coming to their nests, you'll need to arrange a special evening trip, but it is well worth the effort to get close to these enchanting and mysterious birds.

Gannets

Guillemots

Where to watch

Here are some of the best seabird colonies to visit in Britain:

- RSPB Bempton Cliffs
  (cliff-nesting species)
- RSPB South Stack
  (cliff-nesting species)
- RSPB Mousa
  (Storm Petrels)
- Skomer Island
  (Manx Shearwaters)
- RSPB Troup Head
  (cliff-nesting species)
- RSPB Mull of Galloway
  (cliff-nesting species
  including Black Guillemot)
- RSPB Coquet Island
  (terns including Roseate)
- Farne Islands
  (terns and cliff-nesting
  species)

DO IT BY BOAT

Taking a boat trip along the coast past seabird colonies gives you a different (and often closer) view of the action. In the right areas it also gives you the chance to see in daylight those species that are only at their colonies at night. Check harbours along the coast from the colonies, and consider whale-watching trips, as these usually give good views of seabirds as well as cetaceans.

Black Tern

Chlidonias niger

This 'marsh tern' was a breeding bird in the fens of eastern England, and may yet recolonise if suitable habitat is restored. Meanwhile it occurs as a passage migrant in spring and autumn, both inland and offshore.

Length 22–24cm
Wingspan 64–68cm

1	2	3	**4**	5

J F M A M J J A S O N D

How to find

■ **Timing** Black Terns occur in both passage periods, with spring birds most often seen in May, and autumn migrants peaking in August. Their arrival often follows periods of easterly winds.

■ **Habitat** Although Black Terns may be seen overflying almost any habitat, they are most likely to be found feeding over water. Sizeable lakes and reservoirs offer suitable feeding grounds, especially when there has been an emergence of flies, and they will also feed at sea in shallow, sheltered water. They will join other terns and gulls that flock to warm-water outflows, and may be seen travelling along rivers. There is always the chance of a pair nesting in wet fenland – any evidence of breeding should be reported to the local recorder.

■ **Search tips** Visit your local lake or reservoir in spring and autumn, especially during or after easterly winds, and look out for a smallish, square-tailed and dusky-looking tern patrolling quite slowly but tirelessly low over the water, often dropping down to pick items from the surface. Black Terns may also be found (especially in autumn) offshore, often among other terns, though careful observation is required to pick them out from juvenile terns of other species.

WATCHING TIPS

These terns are delightful to watch, and sometimes (especially in spring) turn up in small parties, making a beautiful sight as they wheel and hawk over the water. Spring birds are distinctive but the much whiter autumn birds (juveniles and moulting adults) can be confused with juvenile Common or Arctic Terns. Also, the rarer White-winged Black Tern may arrive in company with Black Terns – pay careful attention to the head and wing pattern to be sure of your identification.

Sandwich Tern

The largest of our breeding terns, the Sandwich Tern is very much a seabird and nests along the south-east and east coasts of England. There are also colonies in north Wales, southern Scotland, and patchily around Northern Ireland and Ireland.

Length 36–41cm
Wingspan 95–105cm

1	2	3	4	5

J F M A M J J A S O N D

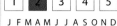

How to find

■ **Timing** This tern is the earliest to return from its wintering grounds, with the main arrival beginning in late March. Most will have departed by early October, although a small number overwinter here.

■ **Habitat** Nesting habitat is on open, shingly or gravelly ground. This may be on beaches or on islands on coastal lagoons, where it will nest alongside Common Terns and Black-headed Gulls. It feeds mainly offshore in shallow and sheltered seas, but may also fish over fresh water next to the coast. Lower Lough Erne in Northern Ireland holds an unusual inland breeding colony, unique in the UK.

■ **Search tips** The loud squeaky-gate call of the Sandwich Tern is often heard before the bird itself comes into view, and adults often call even when flying alone. Among other terns it stands out as a very white and very long-winged bird with a proportionately short tail, and the long black bill gives it a slightly top-heavy look. Look for it fishing offshore, and among other ground-nesting seabirds.

Super sites

1. RSPB Culbin Sands
2. RSPB Lower Lough Erne
3. Lady's Island
4. RSPB Hodbarrow
5. RSPB Coquet Island
6. Cley Marshes WT
7. Pagham Harbour
8. Rye Harbour Nature Reserve
9. RSPB Dungeness

WATCHING TIPS

Sandwich Terns are graceful and feisty birds, interesting to watch both in the air and on the ground. By nesting among gulls they benefit from the gulls' strong defensive reaction to intruding predators, but will readily rise to see off an intruder themselves. When feeding they perform dramatic headlong plunge-dives into the water. In courtship, a male flies high with a fish and calls the female to follow. Look out for overwintering Sandwich Terns – a behaviour that may be on the increase. Winter birds have mostly white crowns and lose their shaggy crests.

Common Tern

Sterna hirundo

This tern is the commonest species in most parts of Britain, and it is the only one that nests well inland as well as on the coast. Inland nesting is most common in south-east and east England, especially where 'tern rafts' are provided.

Length 31–35cm
Wingspan 77–98cm

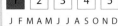

1	2	3	4	5

J F M A M J J A S O N D

How to find

■ **Timing** Common Terns return from their winter quarters during April. Breeding is completed by early August, but migrants can be seen (mostly offshore) into mid-autumn.

■ **Habitat** This tern will nest by fresh water and on undisturbed sea beaches. It uses natural and artificial islands in lakes and will readily take to artificial structures projecting from the water, whether these are purpose-built 'tern rafts' or moored pedalos on a boating lake. It may also nest on shingle banks within slow-flowing rivers. It also fishes over sea and fresh water, including rivers and streams.

■ **Search tips** Common Terns are usually easy to find both inland and offshore. At lakes and reservoirs, scan islands and rafts, even very small ones, for resting birds and look for them flying over the water with relaxed, elastic wingbeats. They can be seen from beaches close to colonies throughout spring and summer and more widely in autumn, often flying low just a few metres out over the water and making dips and dives from prey. They are noisy, especially when breeding.

WATCHING TIPS

Through April and May, Common Tern colonies are full of activity. You may see a courting male offering a fish to his mate, adopting a curious bowing posture as he does so. At this time of year, Common Terns are inclined to take fright for no obvious reason and take off en masse to circle around the colony before returning – unlike with most water birds, this does not necessarily mean a predator is around though a potential nest predator will always elicit a fierce defensive response from the colony. Identification of juvenile and moulting adult terns in autumn can be especially difficult, with Arctic Tern the main confusion species – study the wing patterns to confirm identification.

Sterna dougallii

Roseate Tern

This most elegant of the terns is a very rare breeding bird in Britain, with only a handful of regularly used sites around Northumberland, Ireland and Northern Ireland. It may also be seen offshore in autumn as a passage migrant.

Length 33–38cm
Wingspan 72–80cm

| 1 | 2 | 3 | 4 | 5 |

J F M A M J J A S O N D

How to find

■ **Timing** Roseate Terns tend to arrive later than other species, not occupying their breeding colonies until May. Post-breeding birds disperse around the coastline and then begin to migrate south through August and September, with few seen after the end of September.

■ **Habitat** These terns typically nest on rocky islands or islets, usually in sheltered inshore seawater but occasionally on coastal lagoons. Passage migrants are usually seen offshore, either flying past, feeding around outflows, or resting on offshore rocks.

■ **Search tips** A boat trip off Northumberland to see the Coquet Island birds, or travelling to Lady's Island in County Wexford, is one of the best ways to catch up with this species. Otherwise, sightings are more a matter of chance. Single Roseate Terns may make brief or longer visits to other coastal tern colonies in spring, and in autumn passage migrants can be seen from headlands. Additionally, there are a few sites favoured by post-breeding birds in late summer, but check any gathering of terns through spring and summer as they could hold a Roseate. Beware confusion with the occasional black-billed Common Tern.

Super sites

1. Farne Islands
2. RSPB Coquet Island
3. Hauxley Nature Reserve
4. St Mary's Island
5. Whitburn
6. Lady's Island
7. Dawlish Warren
8. RSPB Dungeness

WATCHING TIPS

Roseate Terns are particularly attractive birds and much sought-after by birdwatchers. Visiting a breeding colony is the best way to become familiar with the species and watch it at length. For those not keen to take a boat around Coquet Island, visit Hauxley Nature Reserve on the Northumberland coast to see the Coquet birds as they come ashore to bathe in the mornings and evenings. It is worth checking tern colonies elsewhere for stray Roseate Terns. Autumn terns should be scrutinised with particular care as it is easy to confuse Common, Arctic and Roseate Terns.

Arctic Tern

Sterna paradisaea

As its name suggests, this tern has a northerly distribution in Britain, with most colonies in Scotland. There are also colonies in north Wales, Ireland, Northern Ireland and the far north of England, nearly all of them at the coast.

Length 33–35cm
Wingspan 75–85cm

1	2	3	4	5

J F M A M J J A S O N D

How to find

■ **Timing** Arctic Terns on northbound migration are seen from mid-April, both along the coast and inland. They may not return to their breeding colonies until early June. Post-breeding dispersal occurs through late summer, and southbound migration through early autumn.

■ **Habitat** These terns form large colonies on open, usually low-lying ground near the coast. They may nest on shingly beaches or on more vegetated ground – grassland or heathland. Inland breeding alongside rivers and freshwater lakes occurs in parts of Scotland and Ireland. Migrants may be seen feeding offshore in sheltered bays, or inland over lakes and reservoirs.

■ **Search tips** At a colony, the birds are noisy, highly active and very evident. Finding them away from their breeding grounds is more of a challenge. Passage migrants may stop off to feed over lakes and reservoirs well inland, but are often overlooked due to their resemblance to Common Terns. The same applies to autumn migrants, which are more likely to be seen offshore. Any passing tern in the migration periods should be scrutinised – the wing pattern is the best identification feature on a bird in flight.

Super sites

1. RSPB Mousa
2. RSPB The Loons and Loch of Banks
3. RSPB Onziebust
4. RSPB Loch of Strathbeg
5. Farne Islands
6. RSPB Coquet Island
7. Lough Corrib
8. Lady's Island Lake

WATCHING TIPS

Visiting an Arctic Tern colony is an exhilarating and slightly dangerous experience. The birds will dive at you if they feel you are too close to their nests, and can draw blood with their bills, so always wear a hat and try to keep a reasonable distance from the nests. You should still be able to enjoy wonderful views, and observe the birds' natural behaviour as they go about the business of nesting. Watching known adult and juvenile Arctic Terns on the nesting grounds will help you to pick them out when you encounter them in other settings.

Uria aalge

Guillemot

The commonest of the four auk species that breed in Britain, the Guillemot nests on steep cliffs around most of the British coastline, except south-east and north-west England. Outside the breeding season it may be seen offshore anywhere.

Length 38–41cm

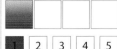

| 1 | 2 | 3 | 4 | 5 |

J F M A M J J A S O N D

How to find

■ **Timing** Guillemots are on their nesting cliffs between March and July, and may be seen on or over the sea near these colonies through spring and summer. In winter they are widely distributed offshore, although few come close enough inshore to be easily observed.

■ **Habitat** This species nests on cliffs and high coastal rocks, open and facing the sea, and can make use of very narrow ledges. Colonies become very large and dense. It feeds on fish caught from a surface dive, so usually hunts on fairly sheltered water, often a long way out from the shore.

■ **Search tips** At cliff-face seabird colonies, Guillemots pack along the most exposed ledges, and are often the commonest species present (or they share the top position with Kittiwakes, which need slightly deeper ledges). Scanning the sea below the cliffs should reveal many birds on the water. Fishing birds will travel some miles along the coast for good fishing grounds – try bays and harbours. When seawatching in autumn and winter, you will probably see many Guillemots flying by or on the sea, although if they are very far out they may be indistinguishable from Razorbills.

Super sites

1. RSPB Troup Head
2. RSPB Fowlsheugh
3. RSPB Mull of Galloway
4. Farne Islands
5. RSPB Bempton Cliffs
6. RSPB Ramsey Island
7. Durlston Country Park

WATCHING TIPS

Guillemots are easy to watch at a cliff colony, although if you are on the clifftop they are unlikely to fly by at eye level. Getting down to sea level within a few miles of a colony may provide closer views of birds on the water, diving for fish. In autumn adult Guillemots may be accompanied by their chicks, which are miniature versions of the adults. Guillemots seen very close inshore in autumn and winter may be sick or oiled – if you find an oiled Guillemot on the beach call the RSPCA or a local wildlife rehabilitation charity.

Razorbill

Alca torda

This is another cliff-nesting auk, which occurs at the same sites as the Guillemot but usually in lower numbers. It may be seen offshore anywhere from the coast outside the breeding season, though again is scarcer than the Guillemot.

Length 37–39cm

1	2	3	4	5

J F M A M J J A S O N D

How to find

■ **Timing** Razorbills occupy their breeding cliffs between March and the end of July. Outside this period, they may be seen offshore.

■ **Habitat** The same cliffs that hold numerous Guillemots are generally home to some Razorbills as well, but the Razorbills do not pack tightly together, and are more likely to use tucked-away crevices, not necessarily facing the sea. They feed in mainly sheltered seas but will tolerate rough conditions quite easily, and spend winter usually some distance offshore.

■ **Search tips** While you will notice the crowds of Guillemots on the cliff face immediately, the Razorbills may require some searching. Check the edges of big Guillemot groups for a pair of Razorbills making use of the cliff where it becomes unsuited to the Guillemots, and scan along the ledges for cracks or fissures that could hold a pair of Razorbills. Scanning the sea below will probably reveal some birds swimming or flying low over the waves. As with Guillemots, Razorbills will be seen regularly from seawatching points in autumn and winter, and feeding Razorbills tend to come closer inshore than do Guillemots (apart from oiled or sick birds).

Super sites

1. RSPB Troup Head
2. RSPB Fowlsheugh
3. RSPB Mull of Galloway
4. Farne Islands
5. RSPB Bempton Cliffs
6. RSPB Ramsey Island
7. Durlston Country Park

WATCHING TIPS

As with Guillemots, Razorbills can be watched at ease at their nests from the clifftops. However, more rewarding eye-level views may be obtained from a boat that goes along the base of the cliffs. Outside the breeding season, the problem of separating the two species can be considerable – try to get a clear impression of the bill shape and head pattern to distinguish them. As with Guillemots, the chicks leap into the sea from their nesting ledges before they can fly, and then are escorted by a parent into the safety of deeper water.

Black Guillemot

This auk is smaller and more compact than Guillemot and Razorbill. It has a restricted distribution in Britain, breeding on coasts around west Scotland, Northern Ireland and Ireland, and remains close to its breeding sites through winter.

Length 30–32cm

How to find

■ **Timing** Black Guillemots can be seen at the same sites all year round, coming to land during the breeding season between mid-spring and mid-summer and spending the rest of its time at sea nearby (and often close inshore). Young birds may disperse some distance, and the likeliest time for a Black Guillemot to turn up away from the usual sites is in mid-autumn.

■ **Habitat** These auks use the lower and less sheer parts of the cliff face for nesting, and will also nest among aggregations of boulders and even on artificial structures. They do not form dense colonies. They fish mainly in quite shallow and sheltered waters.

■ **Search tips** Where Black Guillemots occur on the same cliffs as the other auks, scan lower and 'lumpier' sections of the cliff for them. The red feet and (when the bird opens its bill) scarlet gape are noticeable from a distance. Better views of Black Guillemots can often be had from sea level – when on coastal paths or beaches by sea lochs or open sea, scan the water for swimming birds. Bear in mind that birds in winter plumage look very different to summer adults.

Super sites

1. RSPB Mousa
2. RSPB Dunnet Head
3. Loch Alsh
4. RSPB Mull of Galloway
5. St Bees Head
6. Fedw Fawr

WATCHING TIPS

Black Guillemots are confiding and quite easy to watch at close range, although their chosen habitat means that the best way to do this is from a boat trip. Summer whale-watching trips off the west coast of Scotland offer a great chance of close views. Unlike the larger auks, young Black Guillemots do not stay close to a parent after fledging, but may be seen alone in late summer. Dispersing youngsters may stray well away from breeding grounds and take up residence in a new area for several weeks.

Little Auk

Alle alle

A tiny auk, this species breeds in the high Arctic. It occurs mainly off the north and east coasts of Scotland and north-east England, as a passage migrant. Numbers in UK waters are heavily influenced by weather conditions at sea.

Length 17–19cm

1	2	3	4	5

J F M A M J J A S O N D

How to find

■ **Timing** The classic conditions for a large passage of Little Auks involve easterly gales during November, which drive birds into the North Sea. Extreme conditions can send thousands of birds past seawatching headlands. Prolonged bad weather affects feeding conditions, and weakened, hungry birds may be forced onto beaches or driven inland.

■ **Habitat** This auk is nearly always seen at sea, usually flying past headlands but sometimes also swimming, and occasionally they are beached. They may 'sit out' bad conditions by taking refuge in sheltered bays or river mouths. Storm-driven birds forced well inland often end up in entirely unsuitable situations – in fields and gardens rather than on any form of water, and such birds need to be released at the coast (after any treatment they may need) to stand a chance of survival.

■ **Search tips** Finding this bird is really a matter of watching the weather forecasts and heading for the coast when conditions look right. Because such conditions are often stormy, it is important to position yourself high enough that you can see between the wave crests.

Super sites

1. Fair Isle
2. Fife Ness
3. Farne Islands
4. Hartlepool
5. Flamborough Head
6. Spurn Head
7. RSPB Freiston Shore

WATCHING TIPS

Because the main Little Auk passage period is later than that of other seabirds, those keen to see them will need to make a specific trip for the species and there may be few other birds of interest around. However, a strong migration of Little Auks is quite a spectacular sight. Little Auks are rarely seen outside the late autumn-winter period – many reported sightings in late summer actually relate to juvenile Guillemots or Razorbills, which are much smaller and shorter-billed than adults but do look 'bill-less' like a Little Auk does..

Fratercula arctica

Puffin

Unmistakable and much loved, this is our smallest breeding auk. Its breeding distribution along the coast is similar to that of Razorbill and Guillemot, although it is less numerous at most sites. It is also less often seen in winter.

Length 26–29cm

| 1 | 2 | 3 | 4 | 5 |

J F M A M J J A S O N D

How to find

■ **Timing** Puffins occupy their breeding sites between March and April, and depart in mid-August. Birds may be seen offshore near the breeding grounds throughout this period, and through autumn they disperse more widely. Winter sightings are less frequent as the birds are often well offshore.

■ **Habitat** Uniquely among our auks, the Puffin digs itself a burrow as a nesting chamber. For this it requires sufficiently soft ground, usually near the tops of lower cliffs or on plateaux partway down – colonies also form on the tops of small islets. It will also use rabbit burrows (after driving out the original occupants). The guano encourages the growth of luxuriant grass around colonies. It usually fishes in quite deep water.

■ **Search tips** When at suitable sites, look for Puffins on flat and vividly green grassy spots along the coast. Although during incubation time some birds will be out of sight in their burrows, there will be plenty of activity among the 'off duty' birds. When on the shore near a colony, scan the sea for swimming Puffins, but note that they do not readily come close inshore.

Super sites

1. RSPB Noup Cliffs
2. RSPB Troup Head
3. Isle of May
4. Farne Islands
5. Great Saltee Island
6. RSPB South Stack
7. RSPB Bempton Cliffs

WATCHING TIPS

There are several sites in Britain where you can enjoy the company of Puffins at very close range. Visiting a colony on a boat trip can give more pleasing views than the 'top-down' perspective that you'll have from a clifftop, and will also give you close views of birds swimming and in flight. Visit in mid- to late summer for the chance to see the chicks as they leave the nest and launch themselves into the sea. Pelagic trips into the North Sea in late autumn and winter offer another good opportunity to see and watch Puffins.

Rock Dove

Columba livia

The Rock Dove is a bird of remote coastlines in Scotland, Northern Ireland and Ireland. It is the ancestor of all domestic pigeons, and therefore also the ancestor of the Feral Pigeons that are common in most towns.

Length 31–34cm

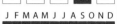

| 1 | 2 | 3 | **4** | 5 |

J F M A M J J A S O N D

How to find

■ **Timing** This species remains in the same areas all year round.
■ **Habitat** Rock Doves breed on remote and rugged cliffs, mainly off north and north-west Scotland (including the islands). Feral Pigeons use building ledges and crannies in towns and cities as a substitute for cliffs. They feed on the ground wherever suitable vegetable matter can be found – for Rock Doves this is in fields near the coast, while Feral Pigeons mainly feed on human leftovers in the street.
■ **Search tips** Feral Pigeons are very easy to find, with a healthy and visible presence in most urban and suburban settings. Flocks usually contain birds in a variety of plumage types including pied, black and 'red' – signs of their domestic ancestry. Finding 'pure' Rock Doves is much more difficult, as some birds of feral origin have returned to a Rock Dove lifestyle, nesting on cliffs and feeding in fields. When in an area that still holds 'pure' Rock Doves, if you find a flock of birds on or near the cliffs that all exhibit the correct Rock Dove plumage pattern, this is as close as you can get to being certain that they are pure wild birds.

Super sites

1. RSPB Handa Island
2. Durness
3. RSPB Dunnet Head
4. Wick

WATCHING TIPS

Feral Pigeons are unpopular with many town-dwellers, but their fearless nature does mean that you can make a detailed study of their behaviour which, like that of most highly social birds, is interesting and complex. You'll see strutting males courting apparently disinterested females, eventually leading to tender mutual preening sessions and other pair-bonding behaviour. Wild Rock Doves behave in exactly the same way, but their shyness and habitat preference means that you're less likely to have the chance to see this behaviour.

Stock Dove

A rather undistinguished, smallish pigeon, the Stock Dove is a bird of woodland and farmland with a wide distribution across Britain except for the far north. It is easily overlooked, and often mistaken for Woodpigeon or Feral Pigeon.

Length 32–34cm

| 1 | 2 | 3 | 4 | 5 |

J F M A M J J A S O N D

How to find

■ **Timing** Stock Doves are present in the same general areas throughout the year.
■ **Habitat** This species feeds mainly on seeds taken from open, sparsely vegetated ground, such as arable and grazed fields and glades and clearings in woods. It will also visit gardens where seed is offered on the ground or on bird tables. It nests in tree holes and hollows so ideal habitat has a combination of stands of mature trees and open ground for feeding.
■ **Search tips** Look for Stock Doves on the ground and high in the branches of tall trees. Areas with scattered large trees can offer easy viewing opportunities as the doves fly between the trees – look for a very plain grey pigeon, lacking the white wing-markings of Woodpigeon and the white underwings of a grey Feral Pigeon. Scan flocks of pigeons feeding on the ground, as Stock Doves may feed alongside Woodpigeons or Feral Pigeons. Males holding territory give a far-carrying, pumping, hoot-like single coo – when you hear this, scan carefully through nearby trees, especially close to the trunk where there may be a nesting hole.

WATCHING TIPS

Stock Doves are often shy and difficult to watch – even birds perched very high in trees will sometimes take fright when being watched. Birds in heavily visited areas such as parks may become more tolerant, but it is unusual to be able to get very close to them. The tree holes in which they nest are often a scarce resource, much sought after by several bird species including Jackdaw, Kestrel and Little Owl – early in the season you could witness conflict between Stock Doves and other species for ownership of a particularly desirable site.

Woodpigeon

Columba palumbus

One of the most common of all British birds, the Woodpigeon can be seen almost everywhere and in most habitat types.

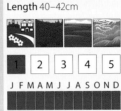

Length 40–42cm

| 1 | 2 | 3 | 4 | 5 |

J F M A M J J A S O N D

How to find it

■ **Timing** Woodpigeons are present in the same areas all year. They form large flocks in winter.

■ **Habitat** Any habitat with some trees and open ground is suitable. The highest numbers are on mixed farmland with scattered woods and copses – they are also common in towns where they will feed alongside Feral Pigeons.

■ **Search tips** Woodpigeons are usually very easy to find. Most gardens will attract them if food is offered on the ground or on bird tables, and any walk in the countryside will produce numerous sightings of birds flying overhead or feeding in fields.

WATCHING TIPS

In town parks, Woodpigeons become as approachable as Feral Pigeons, and may be watched easily at close range. Attract them to your garden by scattering seed on the ground – they may also nest in garden trees. Young birds lack the white neck marking.

Collared Dove

Streptopelia decaocto

This pretty dove first bred in Britain in the 1950s, but rapidly colonised to become a very common and widespread bird.

Length 31–33cm

| 1 | 2 | 3 | 4 | 5 |

J F M A M J J A S O N D

How to find it

■ **Timing** This dove is present in the same areas all year.

■ **Habitat** Most Collared Doves live in the vicinity of towns and villages, and adjoining tree-scattered farmland, although they avoid city centres. They require trees in which to nest, and open ground for feeding. They are common garden birds.

■ **Search tips** Look out for Collared Doves perched on rooftops and telegraph poles, and feeding on the ground around bird tables and feeders. Males holding territory are easily tracked down by their persistent territorial cooing.

WATCHING TIPS

Like most garden birds, Collared Doves can be watched and enjoyed from the kitchen window as they go about their daily lives. Attract them by offering seed on the ground. Territorial males will call from prominent perches, and perform an elegant gliding display flight.

Streptopelia turtur

Turtle Dove

Our smallest and most colourful dove, this species is a summer visitor to England, with its stronghold in the south-east. Migrants turn up elsewhere from time to time. It is a declining species, and can be difficult to find.

Length 26–28cm

1	2	3	4	5

J F M A M J J A S O N D

How to find

■ **Timing** Although the first Turtle Doves return in the second half of April, May is the main arrival month and the easiest time to find the species. As the breeding season progresses, the males sing less often and are thus harder to locate. Dispersing juveniles in September account for most sightings away from the main breeding range.

■ **Habitat** These are birds of quiet open countryside, with areas of trees, hedgerows or bushes for nesting and proximity to fresh water. In early autumn they may appear in less suitable habitat, including (occasionally) in gardens.

■ **Search tips** The easiest way to locate a Turtle Dove is to listen for the soft (but quite far-carrying) purring 'song' of the male in suitable habitat in early May, especially first thing in the morning. The bird itself may be perched in full view on a telegraph pole or in a dead tree, or (especially when newly arrived) more well-hidden in a bush. Turtle Doves frequently come to water to drink, so a vigil in a hide overlooking shallow fresh water may be successful.

Super sites

1. RSPB Middleton Lakes
2. RSPB Minsmere
3. RSPB Stour Estuary
4. Woods Mill Nature Reserve
5. Stodmarsh NNR

WATCHING TIPS

As this species has undergone such an alarming decline, and a contraction in its distribution, it is especially important for birdwatchers to note sightings and any evidence of breeding. One of the main causes of its decline is thought to be reduced supplies of seed (including spilled grain) in the breeding season, so if you live in a rural area where the species occurs, you may be able to help by setting up a feeding area for them. Dispersing juvenile Turtle Doves occasionally take up residence in suburban gardens in the weeks prior to migration.

Ring-necked Parakeet

Psittacula krameri

Native to Asia and Africa, this vivid green parakeet has become established in England following accidental and deliberate releases of captive birds. Its strongholds are in the London area and Thanet in Kent, but it is spreading fast.

Length 38–42cm

| 1 | 2 | **3** | 4 | 5 |

J F M A M J J A S O N D

How to find

■ **Timing** Ring-necked Parakeets can be found in the same areas throughout the year. Very large communal roosts form in the winter months.

■ **Habitat** The species prospers best in habitats with some mature trees but also open areas, although it mainly feeds in trees (on buds and fruit) and requires tree holes for nesting. Extensive parkland can host large numbers. It readily visits gardens and feeds from both bird tables and hanging feeders.

■ **Search tips** These are noisy birds and it is often the raucous, screeching call that alerts you to a bird or birds flying overhead. They can be quite inconspicuous when feeding among green tree foliage, although flocks do tend to keep up a constant chatter of quieter calls when perched and feeding. You may also notice them by the sound of falling food remains, dropped as they eat. Checking medium-sized tree holes (perhaps the old nest-holes of Green or Great Spotted Woodpeckers) in areas where the species occurs is the way to locate nesting birds, although other species use similar-sized holes.

Super sites

1. Kew Garden
2. Hyde Park
3. Richmond Park
4. Lewisham Cemetery
5. Knole Park
6. Margate

WATCHING TIPS

Many Londoners have seen these birds in their gardens, and enjoyed watching them clambering over bird feeders or dextrously using their feet to pass food to their bills. The metal mesh peanut feeders seem especially attractive to the species. They may also use large nestboxes with hole entrances. Recording the spread of Ring-necked Parakeets is important, as their impact on native hole-nesting species is as yet unquantified. Note that green parrots seen away from the core areas could well be other species, or be recent escapee Ring-necked Parakeets rather than established ferals.

Cuculus canorus

Cuckoo

Its call may be the most familiar sound of spring, but many people have never seen a Cuckoo, only heard one. It is very widespread in the UK and Ireland, but is in a period of severe decline, for reasons that are not yet clear.

Length 32–34cm

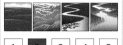

| 1 | 2 | 3 | 4 | 5 |

J F M A M J J A S O N D

How to find

■ **Timing** Most people hear their first Cuckoos in April. With no parental responsibilities, the adults depart as early as July, while juveniles are not ready to begin their migration until August or September.

■ **Habitat** Cuckoos are wide-ranging birds and may be seen in most habitat types. Their presence depends on there being a population of a suitable host species in the area. The main host species in Britain are Reed Warbler, Meadow Pipit and Dunnock, so look for Cuckoos in areas that hold good numbers of these species – around reedbeds, in open grassland and moorland, and around farmland with hedgerows.

■ **Search tips** The 'cuck-oo' call of the male may be given in flight or from a perch. It carries a considerable distance. Check for the calling bird on exposed perches – the tops of hedges, in bare trees, on telegraph wires or on fence posts – look for an elegant, long-tailed shape, often with drooped wings. In flight especially the Cuckoo gives the impression of a small bird of prey, and may cause an alarm or mobbing reaction among small birds that draws attention to its presence.

WATCHING TIPS

The Cuckoo's remarkable life cycle leads to some interesting behaviour. While males are quite showy, females go about their business with great stealth, and few birdwatchers have been lucky enough to witness the moment of a Cuckoo's visit to its host's nest. It is thought that the male helps by distracting the hosts while the female sneaks in to lay her egg. In early summer, if you see a Meadow Pipit or other host species carrying food, watch quietly from a distance to see where it goes and you may be lucky enough to see a juvenile Cuckoo on the nest or nearby, being fed by its diminutive 'parent'.

Barn Owl

Tyto alba

This beautiful owl, high on the lists of many British birdwatchers as 'favourite bird', is widespread although rarely common in most kinds of open countryside throughout Britain. Following steep declines in the last century, its population is being closely monitored.

Length 33–35cm
Wingspan 85–93cm

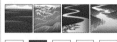
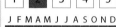

1	2	3	4	5

J F M A M J J A S O N D

How to find

■ **Timing** Adult Barn Owls stay on or close to their breeding grounds all year, and young birds only disperse a short distance. The species is generally easier to see in the long days of summer, where it will begin hunting well before dusk. Weather is also a factor – light winds allow for effective hunting but birds are less active in still or very windy conditions.

■ **Habitat** As a hunter of small rodents, the Barn Owl needs reasonably open ground over which to search for prey – fields (both arable and pasture), roadsides, river banks and marshland are all suitable. It needs a sheltered hollow to nest – in a tree (including single isolated trees) or unused farm building, but it will also use nest boxes.

■ **Search tips** Look for Barn Owls in quiet open countryside in the late afternoon and evening. Scan low over fields, and position yourself with the wind blowing against you as Barn Owls fly into the wind for uplift. Any derelict-looking farm building with an opening for access could attract nesting Barn Owls – watch from a distance to see if any birds come or go.

Super sites

1. RSPB Mersehead
2. RSPB Nene Washes
3. RSPB Strumpshaw Fen
4. RSPB Northward Hill
5. RSPB Elmley Marshes
6. RSPB Pulborough Brooks

WATCHING TIPS

The trick of positioning yourself against the wind relative to a hunting Barn Owl can produce some wonderful close views. If you should be lucky enough to find a Barn Owl nest-site, you could enjoy a summer of priceless experiences, watching the parents bringing prey into the site, and in due course the young birds fledging but remaining nearby as they learn to hunt. If the nest is in an old building, do not be tempted to go inside – watch from a hiding place outside, a good distance away. Note that in rainy and windy weather you may see no activity at all.

Athene noctua

Little Owl

This is not a native species but was introduced to England from mainland Europe in the late 19th century. It has spread to much of England and Wales, and reaches southern Scotland, but is absent from Ireland and Northern Ireland.

Length 21–23cm
Wingspan 54–58cm

| 1 | 2 | 3 | 4 | 5 |

J F M A M J J A S O N D

How to find

■ **Timing** Little Owls are sedentary, staying on their breeding territories throughout their adult lives. They are most active at dawn and dusk, but can often be seen resting at their roosts through the day.

■ **Habitat** With a diverse diet comprised of large insects, worms, small mammals and small birds, the Little Owl favours low-lying, quiet, wildlife-rich habitats with a mixture of woodland, hedgerow and open ground. It needs tree holes (or other suitable hollows, including in buildings) for nesting.

■ **Search tips** In the daytime, Little Owls will sit for long periods in preferred roosting spots, which are often close to the nest-hole. Look for them in large trees, close to the trunk, or tucked into sheltered spots on old farm buildings (for example, in window frames or on ledges under the eaves). If accidentally disturbed, they fly away with a distinctive thrush-like bounding flight, usually retreating to the nearest good-sized tree where they will land and watch you. Alert birds perform vigorous head-bobbing movements, which are noticeable from a distance.

WATCHING TIPS

Once you have located a Little Owl roost site you should be able to enjoy regular views of the birds (usually a mated pair will be present), and the nest-site is probably very close. Little Owls can be nervous though so watch from a good distance – using a car as a hide can help to avoid disturbance. Visiting at dusk or dawn should produce views of the birds hunting – they will commonly pounce on prey on the ground from a low perch, and may pursue beetles and other ground insects on foot. During the day you may see the pair affirming their bond by gently preening each others head and neck plumage.

Little Owl • BRITISH BIRDFINDER **173**

Tawny Owl

Strix aluco

Our commonest owl, the Tawny Owl is found in wooded areas throughout England, Wales and Scotland, but is absent from Ireland and Northern Ireland, as well as some offshore islands. Strictly nocturnal, it is more often heard than seen.

Length 37–39cm
Wingspan 94–104cm

1	2	3	4	5

J F M A M J J A S O N D

How to find

■ **Timing** Tawny Owls are very sedentary, remaining in the same home range all year round and throughout their lives. Young birds in autumn and early winter wander slightly further. The birds are most often heard calling in winter, when they are courting and establishing territorial boundaries.

■ **Habitat** This is a woodland owl, most often found in mature deciduous woodland but also using mixed and sometimes coniferous forest and more open parkland. As it hunts mainly by sound rather than sight it can locate and catch its prey in quite thick ground cover. If there are sufficient trees around it occurs in larger town and city parks and suburban gardens.

■ **Search tips** Most people's experience of this owl is hearing its fluting hoot well after dark in winter – the male's territorial song, sometimes answered by a sharp 'ke-vick' call from the female. On hearing the hoot, if it is a bright night you may see the bird itself – a squat round shape perched on a roof or in a tree or flying on broad wings between trees. Roosting birds may be found by day if you scan treetops near the trunk – they may also be revealed by mobbing songbirds.

WATCHING TIPS

These nocturnal birds are difficult to watch, but clear, bright nights may give good views. Young Tawny Owls leave their nests well before they can fly, and at this 'branching' stage they can be quite easy to see. Occasionally a youngster will fall from the tree, but leave it be unless it is in immediate danger – it is capable of climbing back into the tree. Be careful in the vicinity of a Tawny Owl nest – they have been known to inflict serious injuries on people who have got too close. Most Tawny Owls in Britain are of the familiar 'brown morph' but once in a while you could encounter the striking 'grey morph'.

Long-eared Owl

This beautiful forest owl is widespread in the UK and Ireland, especially the north, but is not especially common. Its strictly nocturnal habits, shy nature and preference for more remote woodlands make it a difficult bird to see.

Length 35–37cm
Wingspan 90–100cm

1	2	3	4	5

J	F	M	A	M	J	J	A	S	O	N	D

How to find

■ **Timing** Long-eared Owls are present all year round, but there is an influx of extra birds from the continent in autumn. Birds that breed in north Scotland move south in winter. In winter communal roosts form in thick scrub.
■ **Habitat** This species breeds in woodland with open ground nearby. On mainland Britain it is usually found in mixed or coniferous forest. There is evidence that it is excluded from richer deciduous woodland by Tawny Owls – in Ireland and Northern Ireland, where there are no Tawnies, it uses all kinds of woodland. Winter roosts form in thick thorny bushes or hedges, while immigrants from the continent in autumn may be seen in quite open coastal habitats.
■ **Search tips** Finding Long-eared Owls in the breeding season is a real challenge. Watching open ground at the woodland's edge at dusk on summer evenings may produce results. Fledged chicks have a loud 'squeaky gate' begging call which can give them away. In winter, look carefully in thick bushes for roosting Long-eared Owls, which may be well hidden. Seawatchers along the east coast may see this species arriving off the sea in broad daylight.

Super sites

1. Pennington Flash
2. Hobhole Drain
3. RSPB Northward Hill
4. RSPB Elmley Marshes
5. Rye Harbour

WATCHING TIPS

Although finding a roosting owl in winter is your best chance of a close view, it may not be very rewarding, as the owl will be doing very little and may be very well concealed. Birds use the same roost site for days or weeks on end but don't always keep to the same spot within their preferred bush, so try another visit and you may get a better view. If you wait nearby until dusk you may see it leave the roost and begin its evening hunt. Locating calling fledglings could provide great views of the adults feeding the brood, but keep a good distance away to avoid disturbance.

Short-eared Owl

Asio flammeus

As the most diurnal of our owls, this is probably the easiest species to see. It breeds mainly on moorland in the northern uplands, becoming more widespread in winter. It is rare in Northern Ireland and Ireland.

Length 37–39cm
Wingspan 95–110cm

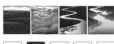

1	2	3	4	5

J F M A M J J A S O N D

How to find

■ **Timing** Most birds will be on their breeding grounds between March and July or August, while the winter range is occupied from October to March, but there is a broad zone of overlap between breeding and wintering grounds and many areas are occupied year-round. Migrants from the continent arrive in autumn.

■ **Habitat** This is an owl of open countryside, primarily rough heathy or grassy moorland in spring and summer, and lower-lying rough pasture, often coastal, in the winter months. Here it hunts by day for Short-tailed Voles, which form the bulk of its diet. Its breeding productivity is much greater when vole numbers are high.

■ **Search tips** Whether you are searching the upland moors in summer or coastal grazing marsh in winter, to find Short-eared Owls scan low across the fields for a long-winged and graceful bird, flying slowly with twists and dips. Scan the ground too, for resting birds, and along lines of fence posts. When seawatching in autumn along the east coast, look out for arrivals coming in off the sea – at some sites dozens arrive in a single morning.

Super sites

1. RSPB Birsay Moors
2. RSPB Trumland
3. RSPB Dee Estuary – Parkgate
4. RSPB Nene Washes
5. RSPB Ouse Washes
6. RSPB Berney Marshes
7. RSPB Elmley Marshes

WATCHING TIPS

Once you find a hunting Short-eared Owl, watch from a concealed position with the wind behind you if possible, and it may come closer. A hunting bird can make a prolonged and thorough exploration of a single small field, allowing for long views. If watching on moorland, be aware that these owls are ground-nesters, and keep to the paths to avoid unnecessary disturbance. It is also important to keep well back from communal roost sites. Watch out for interactions between this species and other open-country birds of prey – aerial challenges and thefts of prey are quite common.

Caprimulgus europaeus

Nightjar

This curious nocturnal bird, related to the Swift, is a summer visitor mainly to lowland heaths in southern England, although it also occurs patchily further north into Wales and just into Scotland. It is absent from Ireland and Northern Ireland.

Length 26–28cm
Wingspan 57–64m

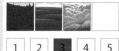
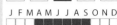

| 1 | 2 | **3** | 4 | 5 |

J F M A M J J A S O N D

How to find

■ **Timing** Nightjars arrive in mid-spring and depart in early autumn, with June and July being the best months to see them as this is when territorial males are at their most active. Migrants occasionally turn up in unexpected places in autumn.

■ **Habitat** These insect-eating birds need open space over which to hunt, and thick ground cover where they can conceal their nests. They tend to be found where there are stands of tall trees adjoining open moor or heath, as the trees boost the numbers of insects and provide perches from which the males can sing. Cleared areas in forestry plantations are often colonised until the new tree growth gets too high.

■ **Search tips** Nightjars are best looked for on warm, still summer nights. Arrive about an hour before dusk and find a likely spot. The churring song of the male begins as the sky darkens, and you may be able to spot the singing bird sitting lengthways on a tree branch. Listen for the sound of a wing-clap, as this indicates the singing bird has taken flight.

Super sites

1. Thetford Forest
2. RSPB Minsmere
3. RSPB Aylesbeare Common
4. RSPB Arne
5. New Forest
6. Ashdown Forest
7. RSPB Broadwater Warren

WATCHING TIPS

Spending an evening on the heath listening to and watching Nightjars is a magical experience. Between bouts of churring, the males will fly with silent grace across the open heath, showing off the white flashes in their wings and tail. They show no fear of people and may well treat you to a very close fly-past. Hunting birds may also be spotted at dawn, but finding a sleeping bird by day takes good luck and great eyesight. When walking in Nightjar country in summer, whether by day or night, keep to footpaths to avoid accidentally damaging nests.

Swift

Apus apus

These scythe-winged birds, with their screaming call and tireless, powerful flight, are a familiar summer sight in towns and also the wider countryside. They are found throughout the British Isles, although they avoid very high ground.

Length 16–17cm
Wingspan 42–48cm

| 1 | 2 | 3 | 4 | 5 |

J F M A M J J A S O N D

How to find

■ **Timing** Swifts begin to arrive through the second half of April, but the main arrival to towns and villages is usually noticed in the first two weeks of May. They tend to fly high in warm fine weather and lower in cooler and windier conditions.
■ **Habitat** Most Swifts that breed in Britain nest in buildings, using crannies or ledges that they can access easily by climbing on sloping or vertical surfaces. As aerial insect-eaters they may feed almost anywhere, and often hunt over open fresh water, hawking low to catch freshly emerged flies.
■ **Search tips** Scan high in the skies on fine evenings in late April to see the first returning Swifts going overhead. Look out for feeding birds over lakes and reservoirs, including those at the coast. They can readily be distinguished from the Swallows and martins that may share their airspace by their all-dark plumage and very long wings. By mid-May, many towns will have an obvious and active Swift population, the birds screaming loudly as they fly fast around the buildings where they nest.

WATCHING TIPS

Swifts are exhilarating to watch. As they only settle when at their nests, much of their social behaviour is conducted on the wing and can be watched from the ground. You can encourage Swifts to nest on your home by adding special nest boxes or 'swift bricks' high on the building. Should you find a Swift on the ground, if it appears otherwise healthy try to 'relaunch' it by holding it high on your flattened hand and moving your hand gently to let it feel the air under its wings. Grounded Swifts are unable to take off on their own as their legs are too small to spring upwards.

Alcedo atthis

Kingfisher

This dazzling small bird is an expert fisher, and is found around all kinds of low-lying open fresh water throughout the UK and Ireland. It is relatively easy to see but rather more difficult to watch at length.

Length 16–17cm

1 2 3 4 5

J F M A M J J A S O N D

How to find

■ **Timing** Kingfishers remain around their breeding area all year round, although young birds may disperse quite widely in autumn. Very cold winter weather forces them to move towards the coast in search of unfrozen water – mortality can be high when there is a prolonged freeze.

■ **Habitat** All kinds of fresh water with a good population of small fish can attract Kingfishers, from small streams to sizeable reservoirs. Sites that are particularly good have clear water, not too deep, with plenty of marginal trees and bushes, providing hunting and resting perches. For nesting they require a vertical bank, well clear of the water line, of soft material in which they can dig their tunnel.

■ **Search tips** Many a riverside walk will yield a glimpse of a Kingfisher, often flying hard and fast out of sight around the next bend. When walking by rivers or lakes, scan ahead above the water line, checking overhanging branches and any suitable perches projecting from the water. If you spot a bird, you may be able to get a closer view with a stealthy approach. Alternatively, just sit and wait quietly. Anglers often enjoy close Kingfisher encounters, just by keeping still.

WATCHING TIPS

Patience and concealment are key to watching Kingfishers. Watching from a hide takes care of the concealment, and there are many nature reserves with hides overlooking water that regularly provide great views of hunting or even nesting Kingfishers. Alternatively, wear dull clothes and find a spot on the lakeside or river bank that offers a good view while allowing you to sit against the bank to conceal your silhouette (avoid doing this anywhere near a nest-site, though). Relaxed Kingfishers may spend long spells on the same perch, giving great views.

Bee-eater

Merops apiaster

This spectacularly colourful bird breeds in the Mediterranean, and is a rare visitor to Britain. It is most likely to be seen in southern regions, and arrivals sometimes involve pairs or small parties. Occasionally they stay to breed.

Length 27–29cm

| 1 | 2 | 3 | 4 | **5** |

J F M A M J J A S O N D

How to find

■ **Timing** The most likely month for Bee-eaters to arrive in Britain is May, when returning migrants may overshoot their regular breeding grounds. Overshoots tend to occur when there is warm and fine weather on the continent, with southerly winds. Arrivals in the autumn passage period do occur but are less frequent, and involve mainly juveniles, dispersing from their breeding grounds.

■ **Habitat** Bee-eaters may be seen flying over any sort of habitat. If they stop to feed it will be a spot with good numbers of large flying insects – low-lying open areas, perhaps near water. Sandpits, quarries and river banks may tempt a pair or pairs to linger and perhaps attempt to breed.

■ **Search tips** It is well worth looking for Bee-eaters and other overshooting species when the weather conditions are right – try open countryside near the south coast. Many Bee-eaters are first detected by call – their distinctive bubbling note is given in flight. Look for a gracefully flying bird with long pointed wings – the silhouette is very distinctive even if light conditions mean no colours can be made out. Bee-eaters will often rest on prominent exposed perches, including overhead wires.

WATCHING TIPS

Keeping an eye on your local bird reports will alert you to arrivals of Bee-eaters, and if there is a significant influx it is worth making several visits to any local sites that might offer suitable nesting conditions. When birds have bred in the past, whenever possible the RSPB has arranged public access as well as protection of the site, and this has given many people a wonderful opportunity to watch the birds. There seems a fair chance of further breeding attempts, although Britain's climate may be too cool to support a viable population.

Upupa epops

Hoopoe

Like the similarly striking Bee-eater, this exotic-looking bird occurs in Britain primarily as an overshooting spring visitor. Most birds are found on open ground close to the south coast. There are also a few visitors in the autumn period.

Length 26–28cm

| 1 | 2 | 3 | 4 | 5 |

J F M A M J J A S O N D

How to find

■ **Timing** Late April to May is the best period to find Hoopoes in Britain. They are most likely to arrive following anticyclonic weather conditions on the continent, with warm weather and southerly breezes encouraging them to continue migrating northwards. Autumn migrants appear in September and October. Neither spring nor autumn birds tend to be long-stayers.

■ **Habitat** Hoopoes feed by searching and probing soft ground for invertebrates, so birds that arrive in Britain tend to seek out lowland fields. Grassy clifftops along the south coast are good places to search in spring, while in autumn they could appear almost anywhere.

■ **Search tips** This bird is showy and eye-catching, so if you are lucky enough to be in the right place at the right time you are unlikely to overlook it. Try walking clifftop paths, scanning the ground ahead and checking fence posts and other potential perches. In flight the bird has a very striking, bold, black and white pattern, and its rounded wings convey the impression of a giant butterfly.

WATCHING TIPS

Hoopoes in Britain are often very approachable and with luck can be watched easily as they feed. They walk slowly, making frequent vigorous stabs at the ground with their long decurved bills. Watch a bird for any length of time and you are likely to see its spectacular crest raised in its full glory. Any summer sightings are particularly notable as they may indicate a breeding attempt – the species has occasionally bred successfully in Britain, and because it does regularly breed just a few miles away on the near continent it could be a potential new colonist.

Wryneck

Jynx torquilla

This unusual small woodpecker used to be a widespread breeding bird in Britain, but now it only occasionally nests in the remote Scottish Highlands. It still occurs in small numbers as a passage migrant, more in autumn than spring.

Length 16–17cm

J F M A M J J A S O N D

How to find

■ **Timing** Wrynecks may be seen in both passage periods. In spring they occur most often on the south and east coasts under typical overshooting conditions, while in autumn most records come from the east coast and follow easterly winds.

■ **Habitat** Although it is a woodpecker, the Wryneck is most often seen feeding on the ground, where it takes ants. It feeds on grassy and bare ground, and will take refuge in bushes if disturbed. Breeding birds nest in tree holes, but unlike the other woodpeckers do not excavate their own holes but use existing ones.

■ **Search tips** Patrolling headlands along the east coast during a spell of autumn easterlies is probably the best approach to finding a Wryneck. Their arrivals often coincide with those of other scarce autumn visitors like Barred Warblers. Walk slowly and quietly, scanning the ground ahead of you, and checking any birds seen moving in bushes. In wooded parts of the central Highlands, where breeding of one or two pairs occurred sporadically in the late 20th century, listen for the ringing, falcon-like call of the male in summer.

Super sites

1. Filey Brigg
2. Flamborough Head
3. Spurn Point
4. Donna Nook
5. Porthgwarra

WATCHING TIPS

Although they are beautifully camouflaged, Wrynecks are not especially difficult to watch and can be very confiding – especially tired migrants, which should always be given plenty of space to feed and regain their strength. Startled birds may freeze rather than run off or fly away, sometimes adopting a curious posture with the head twisted back. Any summer records of this uncommon species, whether in the Highlands or elsewhere, are particularly important, especially if a male is heard singing or any other evidence of breeding is observed.

Picus viridis

Green Woodpecker

This distinctive large woodpecker is common and widespread throughout most of Britain, although not in the uplands of the far north-west, and it is absent from Ireland and Northern Ireland. It is usually seen on the ground.

Length 31–33cm

1	2	3	4	5

J F M A M J J A S O N D

How to find

■ **Timing** Green Woodpeckers can be seen all year round in the same areas.
■ **Habitat** Although they require trees for nesting, these woodpeckers do most of their feeding on the ground, where ants form a major part of their diet. Open grazed (by livestock or wild mammals) grassland with a population of Yellow Meadow Ants is ideal feeding ground – you can tell if the ants are there by the presence of their humped anthills. At least a few trees must be present, but the species is not often seen in dense woodland.
■ **Search tips** Look out for Green Woodpeckers feeding on any area of grassland, from your garden lawn to a woodland glade. They hop on the ground quite unobtrusively, but when flying away the bright yellow rump is eye-catching – the loud laughter-like 'yaffle' call also gives them away. Green Woodpecker nest-holes are large and usually found between 2 and 5 metres above ground in the trunks of sizeable trees. However, in parts of southern England, their nest-holes are often appropriated by Ring-necked Parakeets.

WATCHING TIPS

These woodpeckers are rather timid, and when flushed from the ground will often fly into the safety of a large tree and quickly climb up out of sight. Where they occur in town parks, they can become used to people and are considerably more relaxed about being watched. As favourite feeding sites are visited repeatedly, you may also enjoy success by setting up a portable hide, or simply by finding a good hiding place. In spring you may see the male feeding his mate as part of courtship, while the streaked juveniles appear in June.

Great Spotted Woodpecker

Dendrocopos major

Striking in its black, white and red plumage, this species is our commonest woodpecker. It is found throughout England, Scotland and Wales, and has recently established toeholds in Ireland and Northern Ireland, where it is the only woodpecker.

Length 22–23cm

1	2	3	4	5

J F M A M J J A S O N D

How to find

■ **Timing** Great Spotted Woodpeckers remain around or near their breeding grounds year-round, and are most visible and vocal in late winter and early spring. They become very discreet in the summer when incubating and feeding young chicks.

■ **Habitat** A woodland species, the Great Spotted Woodpecker is found in deciduous, coniferous and mixed forest, and also in more open areas such as parkland and town gardens, provided there are reasonable numbers of mature trees around. It is attracted to garden bird feeders.

■ **Search tips** This woodpecker is very obvious early in the breeding season, with territorial drumming and frequent noisy chases through the trees. At all times, the sharp 'kik' call reveals the presence of a bird – if you hear a single call, look for a rather Starling-shaped bird in bounding flight. Perched birds will also call repeatedly – scan tree trunks and branches, and move around to look from different positions in case the bird is on the far side of a branch. Drumming birds choose resonant dead wood, so when drumming is heard look for the bird on obviously dead branches.

WATCHING TIPS

Great Spotted Woodpeckers are usually quite shy and flighty, but many householders have the privilege of being able to watch them from the kitchen window, visiting the garden bird feeder. Hanging mesh feeders filled with peanuts attract them, as do fat balls and blocks in wire cages. Newly fledged young birds, identifiable by their scarlet caps, appear in June and are often more approachable than their parents. In early spring, pairs are extremely active in establishing and defending their territories, and noisy skirmishes may be observed. To discourage Green Spotted Woodpeckers from raiding garden nestboxes, choose boxes with a metal plate at the entrance, or pick tough 'woodcrete' boxes.

Dendrocopos minor **Lesser Spotted Woodpecker**

This sparrow-sized woodpecker is discreet in its habits and difficult to see. It is also declining rather rapidly, although it can still be found in most good-sized deciduous woodlands in the south-east, as well as some more marginal habitats.

Length 14–15cm

1	2	3	4	5

J F M A M J J A S O N D

How to find

■ **Timing** This woodpecker is best looked for on mild, fairly still mornings in late winter and early spring, when males are drumming and both sexes are more vocal. As spring progresses the birds become quieter, and new leaves on the trees hide their activity.

■ **Habitat** Small and light, the Lesser Spotted Woodpecker can forage among higher and thinner twigs than our other woodpeckers. It favours mature deciduous woodland but may also be found in parkland and gardens. In winter, wandering birds can turn up in reedbeds and scrubland, although rarely far from more conventional woodland habitat.

■ **Search tips** Move slowly through suitable woodland, listening for calls and drumming, and pause often to scan the tops of trees with binoculars. The bird's style of movement is unobtrusive, as it runs rapidly up vertical branches, slipping quickly from branch to branch. Its size and preferred foraging height means that even its silhouetted shape is often not obvious. So search patiently, especially if any likely sounds (including tapping) are heard. Nest-holes are small (about the size of the hole on a typical tit nest box) and may be in branches rather than trunks – observe a potential nest-hole from a good distance.

WATCHING TIPS

Lesser Spotted Woodpeckers are not so much shy as unobtrusive. Once you have found a male holding territory or resident pair, repeated observations in the same area should be possible, although will become more difficult once the trees come into leaf. Winter birds foraging at low altitude may come quite close to a patient observer. If there are Lesser Spotted Woodpeckers living near your garden, you may attract them to feed by positioning a fat block in a hanging cage-type feeder high in the branches.

Woodlark

Lullula arborea

This is a rather small and short-tailed lark with a beautiful song, found mainly on lowland heaths in southern and eastern England. Some eastern birds move away in the winter, but southern and south-western populations are resident.

Length 15cm

1	2	3	4	5

J F M A M J J A S O N D

How to find

■ **Timing** Woodlarks are mainly quite sedentary, although numbers in the Norfolk Brecklands go down in the winter months. They are easiest to find in spring, when the males are singing most. Small flocks form in winter.

■ **Habitat** This species needs dry open habitats, ideally with a mix of grassland (better for feeding) and thicker ground cover like heather (better for nesting). It also needs some elevated perches in the form of bushes or posts, from which the males will sing. It will colonise newly cleared areas within forestry, although it will move on when new trees grow up too high.

■ **Search tips** This is a rather inconspicuous bird, especially when quietly feeding on the ground. Singing males are easier to locate as they usually sing from the top of a bush or hillock, and sometimes in a short songflight. The song is very melodious, steady-paced and descends the scale. Males may sing at any time of year but spring is the peak time. If no song can be heard, carefully scan the ground for feeding birds.

Super sites

1. Weeting Heath
2. RSPB Minsmere
3. RSPB North Warren
4. RSPB Arne
5. New Forest
6. RSPB Farnham Heath
7. Thursley Common

WATCHING TIPS

The Woodlark is much admired for its song, and is well worth seeking out just to enjoy its lovely voice. Visit a suitable site in mid-spring and take your time walking around, listening and looking for the birds. The songflight, especially in males that are yet to find a mate, is quite dramatic with a spiralling ascent to around 100 metres, followed by a series of irregular dipping circles and then a rapid descent with wings closed. Birds seen away from the usual sites should be reported as it may be possible to enhance the habitat for them.

Alauda arvensis

Skylark

This lark with its iconic towering songflight is one of the most widespread British birds, and can be found in open countryside throughout the British Isles. Nevertheless, it has declined dramatically in the last century, especially in farmland.

Length 18–19cm

| 1 | 2 | 3 | 4 | 5 |

J F M A M J J A S O N D

How to find

■ **Timing** Skylarks can be found in open countryside year-round, although the birds in the uplands of northern Scotland, northern England and central Wales move to lower ground for the winter, and some continental birds may also visit in winter.

■ **Habitat** Skylarks use all kinds of open landscapes, from low-lying farmland to rugged upland moors. They need a reasonable amount of ground cover in which to nest – enough to conceal the nest and incubating parent. They will use meadows, heathland, grassy moorland and arable fields.

■ **Search tips** The distinctive song is usually the first sign that a Skylark is around. However, hearing the song does not necessarily mean you will easily see the bird – it may have ascended so high that it is almost lost to view. Scan the sky for the speck of a bird – eventually it will descend and give you a better view. Males will also sing from the ground or a raised perch, such as a fence post. Flocks form in winter – scan the ground for feeding birds, and listen for the chirruping flight calls of a flock on the move.

WATCHING TIPS

Throughout a spring or summer day, a male Skylark will perform its songflight several times, and may also get involved in territorial chases with neighbouring birds where the population density is high. Watch patiently and you may see some interesting behaviour. Be very careful when walking in open countryside and if you see agitated Skylarks around retrace your steps as you may be near a nest. In autumn, listen for the rolling call as dispersing birds fly overhead – this 'visible migration' is best observed from hilltops. In very cold weather, Skylarks occasionally visit country gardens, where they will take seed from the ground.

WOODLAND BIRDWATCHING

Before the advent of wide-scale farming, Britain would have been almost entirely forested, with deciduous woodland in the south and Caledonian pine forest in the Scottish Highlands. We may have lost much of our woodland since then, but our bird life still reflects those times, and we still have some wonderfully rich woodlands that are full of a diversity of birds.

The wood and the trees

Birdwatching in woodlands is more challenging than looking for birds in open countryside, simply because the view is so much more obscured. A flying bird can be lost within the maze of branches in an instant, never to be seen again. Watching birds feeding in the high treetops is a literal pain in the neck. You can be listening to a song or calls for hours on end and still not manage to get a clear view of the bird itself.

However, there are a number of things you can do to improve your chances of seeing more birds and getting better views. Perhaps the most important is time of year. Trying to find birds in deciduous woodland in high summer is very difficult – the trees will be in full leaf, and the birds themselves will have more or less stopped singing. A woodland in midsummer can seem almost dead, as far as birds are concerned.

Winter is a better time. The leafless branches provide fewer hiding places for the birds, and many species will be roaming widely in mixed flocks, moving quickly from one feeding spot to the next. Perhaps better still is early to mid-spring, before the leaves have fully grown in, but after the summer visitors have begun to arrive, and all the birds are singing. This is by far your best chance to catch up with elusive species like Lesser Spotted Woodpecker, and to get good views of summer migrants like Redstart and Wood Warbler. In coniferous woodland, there are leaves all year, but the pattern of bird activity is the same as in deciduous woodland.

Location, location, location

When birdwatching in woodland, you will often have more luck by sitting and quietly waiting, rather than walking about. Birds quickly become quite relaxed in the presence of a still, silent human, especially if you wear dull colours or even camouflaged clothing. If you really want to try for the long haul you could set up a hide (having obtained the landowner's permission first). However you go about it, the sit-and-wait tactic will work best if you pick the perfect location.

To choose a good spot, consider the sorts of resources that birds will be seeking. Fresh water is especially important for seed-eaters, and for birds of prey, which bathe often. Woodcocks also frequent areas with soft, wet ground. Woodpeckers are attracted to rotting wood, either standing or fallen. Beechmast is an important food supply for finches and tits in autumn, while Jays are avid collectors of acorns. If you choose a sunny spot near a flowery glade, there will be more flying insects about which attract flycatchers and other insect-eaters.

Great Spotted Woodpecker

Woodcock

Where to watch

Here are some of the best woodlands for birdwatching in Britain:

- RSPB Nagshead
- RSPB Garston Woods
- RSPB Blean Woods
- Gwenf RSPB frwyd-Dinas
- RSPB Coombes and Churnet
- RSPB Wood of Cree
- RSPB Abernethy Forest
- Forest of Dean
- Wyre Forest
- New Forest

THE WOODLAND IN YOUR GARDEN

We may have lost a very significant part of our forest cover, but gardens go a long way towards replacing it. The majority of our familiar garden birds are really woodland birds, which find gardens an acceptable alternative to their natural habitat. By making your garden more like a woodland, you will help support them and the other species on which they depend.

Preserve native trees in your garden wherever possible, and choose native species for new plantings. Provide nest boxes for those species that naturally nest in holes in trees. Decaying wood may not be pretty but it supports a great abundance of invertebrate life that in turn sustains birds, so consider adding a woodpile in a shady spot.

Tree Pipit

Shore Lark

Eremophila alpestris

This attractive lark breeds in the European Arctic and migrates south for winter. Variable numbers – a few hundred in the best years – overwinter in Britain, mainly along coastlines (especially the east coast) but occasionally inland in Scotland.

Length 14–17cm

1	2	3	4	5

J F M A M J J A S O N D

How to find

■ **Timing** Shore Larks are long-staying winter visitors, arriving as early as October and lingering well into April before departing for the short Arctic breeding season. Passage migrants may be seen in Scotland in May. Many remain in the same general area throughout their stay.

■ **Habitat** Typical wintering habitat is quiet shingle beaches with some vegetation, and also coastal stubble fields, saltmarsh and dunes. Here they may join with Skylark flocks, and may also associate with other shoreline passerines like Snow Buntings, Twites or Linnets.

■ **Search tips** Although this species looks striking and colourful in illustrations, its pattern helps it to disappear when feeding quietly on shingly ground. Particular care is therefore needed when scanning – go slowly and watch for sudden movement. If disturbed, it gives a long-winged impression, and frequently gives a high, sibilant two-note call as it flies away. Although individuals are quite loyal to a general area, they can be wide-ranging along a stretch of coastline, and you may require patience and perhaps repeated visits to catch up with them. However, they are also fairly confiding and so if you keep still and wait they may come to you.

Super sites

1. Saltfleetby NNR
2. RSPB Titchwell
3. Holkham
4. Cley Marshes WT
5. Salthouse
6. RSPB Elmley Marshes
7. Rye Harbour Nature Reserve
8. RSPB Dungeness

WATCHING TIPS

With patience these birds will give good views, and you can watch them as they feed, working over the ground systematically and moving with short dashes in between inconspicuous shuffling. The occasional pair has bred in Britain, in the Scottish Highlands, but the prospects for further breeding look rather poor, given that the core Scandinavian breeding population is declining. However, any sightings of Shore Lark in the summer months could indicate a breeding attempt so should be reported to the local recorder.

Riparia riparia

Sand Martin

This is the smallest member of the swallow family to breed in Britain, and the one most closely associated with water. It can be seen across the whole of the UK and Ireland throughout spring, summer and early autumn.

Length 12cm

| 1 | 2 | 3 | 4 | 5 |

J F M A M J J A S O N D

How to find

■ **Timing** Among our summer visitors, Sand Martins are noted for being early returners, with the first sightings sometimes at the end of February. Their arrivals usually precede Swallow and House Martin by a couple of weeks. They are on their breeding grounds between mid-March and August but are more widespread before and after this period.

■ **Habitat** Sand Martins nest colonially in tunnels dug into soft banksides, usually adjoining (or near to) fresh water. Tall river banks, quarries, gravel pits and bare sandy banks and cliffs are all suitable – they will also use artificial purpose-built Sand Martin walls. They hunt (often in company with House Martins and Swallows) for aerial insects, mainly low over open water but also over fields or on fine days high in the sky. In autumn large flocks form, often roosting in reedbeds.

■ **Search tips** Almost every lowland lake or reservoir will attract newly arrived Sand Martins in early spring, and again in autumn after breeding when the adults and young birds disperse. In between, search anywhere there are suitable banksides. Look for a small, 'narrow-looking' and compact swallow-like bird, and listen for the very dry-sounding brief flight call.

WATCHING TIPS

Watching the comings and goings at a Sand Martin colony is a pleasant way to while away a sunny afternoon. Be careful not to get too close though – if the birds begin calling more frequently and stop visiting their nest-holes, move back and give them more space. If watching a riverside colony, position yourself inconspicuously on the opposite bank if possible. In autumn, this species is often seen flocking with its close relatives, allowing you the opportunity to compare adults and juveniles of all three species side by side.

Swallow

Hirundo rustica

The much loved harbinger of summer in the countryside, the Swallow is a common and distinctive breeding bird across rural areas throughout the whole of the UK and Ireland, returning to the same nest-site year after year.

Length 17–19cm

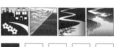

| 1 | 2 | 3 | 4 | 5 |

J F M A M J J A S O N D

How to find

■ **Timing** Swallows begin to return in March, joining Sand Martins to feed up over reservoirs and other open water. At nest-sites, most returning birds are noticed in April. Breeding is completed by September (occasionally early October) and flocks form at inland waters and on the coast before migration. There are very occasional winter records.

■ **Habitat** This species is called Barn Swallow in North America, indicating its preferred nest-site. Any unused or little used rural building with access (a broken window is sufficient) and suitable nesting ledges will be colonised. It hunts over all kinds of ground, especially open water and grazed pasture, both of which offer a good supply of flying insects. Migrants in spring and autumn may be seen arriving or departing along the coasts.

■ **Search tips** Swallows are usually much in evidence when you are in the general countryside – perched on overhead wires, skimming low over the fields or singing from telegraph poles. Before and after the breeding season is the most likely time to find them well away from usual sites.

WATCHING TIPS

If you live in a rural area, you may already have Swallows nesting in your property. Making sure access is not restricted and disturbance kept to a minimum should ensure that they return year after year, and you should enjoy wonderful views of the parents tending to their growing families. In dry springs, keep a patch of ground wet so mud is available for nest-building. If you watch for 'visible migration' from hilltops on autumn mornings, you should see many Swallows, streaming purposefully towards the coast. There are few better ways to spend a sunny summer afternoon than watching Swallows hawking over a flowery meadow.

Delichon urbicum

House Martin

This is the member of the swallow family that you are most likely to find in towns, although it also breeds in rural areas. It is common and widespread throughout the UK and Ireland, except for the extreme north-west of Scotland.

Length 12.5cm

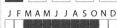

1	2	3	4	5

J F M A M J J A S O N D

How to find

■ **Timing** House Martins usually arrive slightly later than Sand Martins and Swallows, first appearing in late March but not noted at its nest-sites until mid-April. The breeding season can extend well into autumn, with the last broods only fledging in October, not long before most birds have departed for Africa..

■ **Habitat** This species nests mainly on the outside walls of buildings, constructing a mud nest right under the eaves. There are also a few cliff-nesting colonies. It feeds over all kinds of habitats, especially open water, but also above town buildings. Pre-migration flocks form by water and at the coast.

■ **Search tips** Smaller and quieter suburbs, towns and villages tend to support more House Martins compared to larger towns and cities. Look for the unmistakable semicircular mud nests – very suitable well-protected eaves will have a whole row of them – and in spring look around muddy puddles for House Martins collecting mud. When watching a mixed group of hirundines (birds of the swallow family) feeding over open water, look out for the House Martin's square white rump, which immediately stands out among dark-backed Swallows and Sand Martins.

WATCHING TIPS

You may attract House Martins to nest on your own home by providing special House Martin nest boxes. If these are occupied, other pairs may be attracted to build their own nests alongside the artificial ones. Making sure wet mud is available nearby for nest-building may make all the difference. This species has declined somewhat, so any help that you can give it will be valuable. A pair may rear three broods in a summer, with third broods sometimes fledging as late as early October.

Tree Pipit

Anthus trivialis

This pipit is very similar to the Meadow Pipit, but prefers more wooded habitat, and is only here as a summer visitor. It is quite widespread in England, Wales and Scotland but does not breed in Ireland or Northern Ireland.

Length 15cm

1	**2**	3	4	5

J F M A M J J A S O N D

How to find

■ **Timing** Tree Pipits first arrive on their breeding grounds in early April, and most have departed for their African wintering grounds by mid-September. You are most likely to see the songflight in spring, in the early part of the breeding season

■ **Habitat** Although it is still associated with somewhat open habitats and avoids dense, continuous woodland, the Tree Pipit is often associated with woodland edges, such as cleared areas within conifer plantations, or the edges of birch woodland adjoining moorland, and isolated stands of pines within lowland heath. However, it mostly feeds on the ground rather than in trees. It is most numerous in the northern and western parts of its range.

■ **Search tips** This bird performs a distinctive songflight, and hearing the song should result in good views once you find the bird as it will repeat the songflight twice or more in succession. Listen for a quite varied, powerful and continuous series of high-pitched whistles and trills, and look out for the singing bird parachuting slowly earthwards. The flight call is distinctive and powerful too. Scan overhead wires and the tops of bushes and small trees for perched Tree Pipits.

WATCHING TIPS

The Tree Pipit and Meadow Pipit are very similar and there is some overlap in their habitat preferences. Therefore, identification is the first priority. If you hear the song or calls you can generally make a confident identification but if not, it comes down to subtle differences in structure and plumage. In good Tree Pipit country there should be many opportunities to watch and enjoy the songflight, and carefully watching the singing bird should help you to become familiar with the ways its 'look' differs from the more common Meadow Pipit.

Anthus pratensis

Meadow Pipit

In many remote open areas this is by far the most frequently seen small bird. It occurs commonly throughout the UK and Ireland in all kinds of open countryside, although it abandons the very highest ground in winter.

Length 14.5cm

1	2	3	4	5

J F M A M J J A S O N D

How to find

■ **Timing** Meadow Pipits are with us throughout the year. Birds that nest on the highest hills will migrate short or longer distances to lower ground after breeding, some moving to the continent, and large numbers may be seen going overhead from hilltops in autumn..

■ **Habitat** This is a bird of open habitat, including farmland, rough grassland, moor, heath and (especially in winter) saltmarsh. It feeds mainly on the ground but surveys its habitat from visible perches, such as gates, fence posts and hummocks of ground. In very cold weather it disperses widely in search of feeding grounds and will occasionally visit gardens.

■ **Search tips** You will probably 'find' a few Meadow Pipits while walking in the countryside by flushing them from the ground where they were feeding unseen – they fly away with a distinctive thin 'zweet zweet zweet' call. Singing males draw attention when they perform their rising and then parachuting songflight – both song and flight pattern is much simpler than that of the Tree Pipit. Scan elevated points for perched Meadow Pipits, and watch (and listen) for them migrating overhead in autumn.

WATCHING TIPS

As this is the most common British pipit, and also is rather variable depending on its state of moult, becoming familiar with it will help you to notice when you find a different pipit species. Any area with a high breeding population of Meadow Pipits will attract Cuckoos in spring, and the open nature of the habitat means that with patience you could enjoy great views of the two species interacting. If Meadow Pipits come to your garden in a winter freeze, you can help them survive by putting out mealworms or fat-based foods on the ground.

Rock Pipit

Anthus petrosus

Rock Pipits are bigger and stockier than Meadow Pipits, and are found almost exclusively on rocky coastlines. They can be found around our entire coastline, although numbers thin out significantly in the south-east, and also around north-west England.

Length 16.5–17cm

1	2	3	4	5

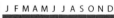

J F M A M J J A S O N D

How to find

■ **Timing** Rock Pipits remain around their breeding grounds throughout the year. The state of the tide will influence exactly where on the beach or foreshore they forage.

■ **Habitat** These pipits can be found on reasonably sheltered rocky shorelines and low cliffs, feeding among the rocks and around artificial structures like piers and groynes, and even wading in the shallows. They may also come onto roadsides and promenades adjoining the beach, on coastal grassland, and on grassland at the tops of low cliffs. The Scandinavian subspecies is occasionally reported in winter from inland sites, such as reservoir foreshores.

■ **Search tips** Whenever you are on a beach with rocks or cliffs nearby you stand a good chance of seeing a Rock Pipit. Look for a smallish, dark and rather featureless bird walking or flitting around the rocks or the base of cliffs. The call is similar to that of the Meadow Pipit but stronger. Unlike Meadow Pipit it rarely forms flocks, but one or two may feed with a Meadow Pipit flock on coastal grassland.

WATCHING TIPS

Rock Pipits can be very tolerant of people, especially when they live on rocky shorelines around small seaside towns. Visit at high tide where their available foraging habitat is more restricted and you could watch them feeding at close quarters, and, in mid-summer, feeding their newly fledged chicks around the rocks. Look out for the Meadow Pipit-like songflight in spring. Rock Pipits are sometimes used as hosts by Cuckoos and, as with Meadow Pipit, the open habitat means that interactions between the two are observed relatively often.

Water Pipit

This pipit, which until relatively recently was considered a subspecies of Rock Pipit, does not breed in Britain but is a scarce winter visitor, mainly to the south and east of England, and often visits the same sites year after year.

Length 17–17.5cm

| 1 | 2 | 3 | **4** | 5 |

| J | F | M | A | M | J | J | A | S | O | N | D |

How to find

■ **Timing** Water Pipits are winter visitors that arrive early and depart late; they are mainly present between October and April but are occasionally seen as early as August and as late as May. By mid-spring they may be more difficult to find but by way of compensation will be in their attractive and distinctive pink-flushed breeding plumage.

■ **Habitat** Most Water Pipits in Britain are seen in low-lying habitats, on quite lush ground around shallow fresh water – boggy or flooded pasture, the edges of marshland, riversides, and well-vegetated lake shores. They are not generally found around salty or brackish water, although they may use marshes that are occasionally flooded by very high tides.

■ **Search tips** The best way to find a Water Pipit is to visit one of the regular sites that attract it, and scan carefully and patiently for birds feeding on the ground. They may be in company with Meadow Pipits, so examine each bird carefully. Water Pipits can be remarkably sedentary in winter, patrolling the same few square metres every day, so if you hear of a sighting, even a few days later, it is worth finding out and checking the exact spot.

Super sites

1. Ribble Estuary
2. RSPB Lakenheath Fen
3. London WWT
4. RSPB Rainham Marshes
5. Stodmarsh NNR
6. Filsham Reedbeds

WATCHING TIPS

Distinguishing this pipit from Meadow Pipit, with which it may flock, requires a good view. Look out for a larger and greyer pipit with a more strongly marked face and, in spring, a lovely rosy flush to the breast and belly. Another identification pitfall is the Scandinavian subspecies of Rock Pipit, which looks rather similar but does not have pure white outer tail feathers. If you are lucky enough to live near a good Water Pipit site, visit often to become familiar with the birds' daily habits and work out how to get the best views.

Yellow Wagtail

Motacilla flava

This is our smallest wagtail and the only one that is a summer visitor. It is found patchily across most of England and Wales except the far west, and the south and south-west of Scotland, but not Ireland or Northern Ireland.

Length 17cm

1	2	3	4	5

J F M A M J J A S O N D

How to find

■ **Timing** The main Yellow Wagtail arrival is in the first half of April, although some are seen in March. They depart from breeding areas in September, and may be seen on passage on the coasts in early autumn.

■ **Habitat** Look out along open flat coastlines for newly arrived migrants in spring, before they spread inland to their breeding grounds. Pastures and meadows with plenty of insects are ideal for this species. Fields used for grazing cows and horses hold large numbers of the flies they feed on, in fact horse manure is used on some nature reserves to attract Yellow Wagtails. Later in the season they may roost communally in wheat fields.

■ **Search tips** These lovely wagtails can become almost invisible when feeding on the ground among longish green grass and flowers. Singing males pick elevated spots such as fence posts or isolated tall plants and are then quite obvious. Also look out for Yellow Wagtails closely following grazing livestock, ready to seize any insects that are disturbed by the animals as they move around.

WATCHING TIPS

This is an intriguing species, with a remarkably large number of subspecies across Europe and into Asia, of which six have been recorded in Britain besides our regular subspecies (*M. f. flavissima*). It is well worth checking through flocks in autumn for any oddities, while the Blue-headed subspecies from the near continent does occasionally breed here. The rare Citrine Wagtail also looks similar to the Yellow Wagtail. Yellow Wagtails are declining, so it is important to inform local recorders of their presence on a new site, or disappearance from a previously used site.

Motacilla cinerea

Grey Wagtail

This very attractive wagtail has quite specific habitat needs but is nonetheless common and widespread throughout the UK. It is a summer visitor to the northern uplands, and tends to visit southern towns only in winter, but is otherwise resident.

Length 18–19cm

1	2	3	4	5

J F M A M J J A S O N D

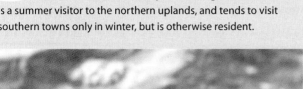

How to find

■ **Timing** You can see Grey Wagtails in most of Britain all year, but those on more northerly and high ground move south and to lower altitudes in winter, while in the south and east they are noticeably more widespread in winter – they also visit gardens far more in winter.

■ **Habitat** The single factor that practically guarantees the presence of Grey Wagtails is shallow, fast-flowing fresh water. Rocky upland streams support them, and so do weirs and locks on canals, slower rivers and dams at the edges of reservoirs. They hunt the small invertebrates that thrive in this kind of water. In winter particularly they will also visit towns (often feeding around puddles on flat rooftops) and gardens, and forage around the muddy margins of lakes and ponds.

■ **Search tips** These birds are showy and eye-catching, the constant bobbing of their long tails drawing the eye as much as the vivid yellow undertail when they take flight. Any time spent around suitable habitat should produce sightings. Scan the edges of streams, and any rocks or other structures projecting from the water, and listen for the shrill, ringing flight call.

WATCHING TIPS

Grey Wagtails show little fear of people and can be watched feeding and even delivering food to their fledged chicks at fairly close range. Males will sing from prominent perches but also perform a Tree Pipit-like songflight. Grey Wagtails are most likely to turn up in gardens that are near streams or rivers, or have a good-sized pond, but in very severe weather may appear in more unlikely settings in the search for food. They will eat mealworms and may take fat-based bird food pellets. They may also make use of open-fronted nestboxes.

Pied/White Wagtail *Motacilla alba yarrellii/alba*

The Pied Wagtail is the British subspecies of the continental White Wagtail. Pied Wagtails are very common and widespread residents to the UK and Ireland, while White Wagtails occur mainly as passage migrants in the autumn months.

Length 18cm

| 1 | 2 | 3 | 4 | 5 |

J F M A M J J A S O N D

How to find

■ **Timing** Pied Wagtails are easy to find all year round, although are much more common in urban environments in the winter months than in the summer. White Wagtails are mainly found between August and October.

■ **Habitat** This insect eater forages mainly on the ground in open habitats. Most breed on or around farmland, often nesting in crevices in farm buildings (they will also use open-fronted nest boxes). They forage in low pasture, on bare ground and around still and flowing fresh water, in flocks during the winter months. In towns, they feed in gardens and on roads and pavements.

■ **Search tips** Pied Wagtails are striking and easily to spot, with their loud two-note flight calls and conspicuous undulating flight. Look out for them on the ground, moving with a rapid running gait, and around water. Like other wagtails they constantly bob their tails while feeding, a movement that draws the eye. In winter they roost communally, often in towns, and are very vocal as they arrive at the roost. White Wagtails, which are paler than Pied Wagtails, are most often found along the east coast, sometimes with Pied Wagtails.

WATCHING TIPS

These charming birds are easy to watch as they are usually fairly unworried by people, although as with all birds you should give them plenty of space when they are nesting. Town birds may exhibit interesting behaviour in winter – they have been noted picking squashed insects from the tyre treads and bumpers of parked cars, and the communal roosts (often in small ornamental trees in town centres) are fascinating to watch. Locate the roost by looking for arriving birds, which often perch and call from rooftops before flying down to settle in the branches.

Bombycilla garrulus

Waxwing

This spectacular and beautiful bird is a winter visitor from Scandinavia and Russia. Tens of thousands visit in irruption years (when a poor berry crop on the continent forces them to range further) but only a few hundred at other times.

Length 18cm
Wingspan 110cm

| 1 | 2 | **3** | 4 | 5 |

J F M A M J J A S O N D

How to find

■ **Timing** The earliest Waxwings appear in mid-autumn, along the north-east coasts. As autumn progresses, they spread further south and west, although it is only in exceptional years that any numbers reach southern Ireland and south-west England.

■ **Habitat** When Waxwings come to Britain they are in search of berries, and gather to feed in areas where there are plenty in close proximity. This means that the highest numbers occur in town environments, especially shopping centre car parks, where numerous berry-bearing ornamental trees are planted. They will rest in nearby taller trees in between spells of feeding, and may be seen overflying most other habitats.

■ **Search tips** It is usually evident by November whether we are having a good Waxwing winter, with birds more likely to be widespread. In any case, visiting towns near the east coast where there are good stands of berry-laden ornamental bushes or trees gives you the best chance of success. Look for stocky, Starling-sized birds in flocks in berry bushes or bare treetops. The pointed crest is obvious from a distance. Also listen for the high trilling call.

WATCHING TIPS

In a good year, such as the 2010–2011 winter, you can enjoy wonderful close views of Waxwings for months on end without travelling far from home. The birds descend in dense flocks upon berry trees and rapidly consume the entire crop, before moving on to the next tree, and show no interest in people watching them just a few metres away. As winter progresses, the larger flocks break up into smaller parties, wandering widely in search of more food. Plant suitable trees (white-berried rowans seem to particularly appeal to them) and they could even visit your garden to feed.

Dipper

Cinclus cinclus

The Dipper is a remarkable bird of fast-flowing rivers, superficially like a thrush but in its way of life quite unlike any other British bird. It is most common in northern and western uplands, and is absent from south-east England.

Length 18cm

| 1 | 2 | 3 | 4 | 5 |

J F M A M J J A S O N D

How to find

■ **Timing** Most Dippers remain on their breeding grounds all year round, although those on the highest and most northerly streams may move downstream in the coldest months.

■ **Habitat** Only fast-flowing shallow and rocky rivers and streams are suitable habitat for this species, which feeds almost exclusively by and in the flowing water. For nesting it needs crevices clear of the water flow (though perhaps only just), among boulders or within artificial waterside structures like bridges. They are not unduly worried by proximity to people and may breed on quite urban stretches of suitable rivers provided there are sufficient aquatic insects to eat.

■ **Search tips** Along suitable streams and rivers, keep an eye out for a Dipper flying by, looking very round and short-tailed and skimming just over the water. Also scan ahead, checking rocks by or within the water flow. The song, given by both males and females, is sweet, melodious and high-pitched, audible over the noise of the rushing water, while startled birds give a succession of short 'zit' calls as they fly away.

WATCHING TIPS

This very specialised bird is most interesting to watch, and using binoculars from a good viewpoint you should be able to watch its feeding behaviour under water as well as above the surface. It immerses itself fully, either dropping in from flight or walking into the water, and walks along the river bed, steadying itself with its wings and keeping its place by strongly gripping underwater stones as it searches for the larvae of aquatic insects. On surfacing it may float or swim downstream to a convenient exit point. Juveniles have a very striking scalloped plumage.

Troglodytes troglodytes — **Wren**

This very small bird is common and widespread throughout the UK and Ireland, and will visit most gardens.

Length 9–10cm

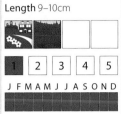

| 1 | 2 | 3 | 4 | 5 |

J F M A M J J A S O N D

How to find it

■ **Timing** Wrens can be seen in suitable habitat all year, but are perhaps easiest to see in spring when males are singing frequently to establish their territories.

■ **Habitat** Anywhere with trees or bushes and a dense understorey will hold Wrens – they will breed in gardens, woodland, scrub and along hedgerows. Thorny bushes offer suitable nesting sites.

■ **Search tips** Listen for the loud, rattling song and scan around to find the singing Wren. In spring they will sing from more exposed and elevated perches. Look on the ground for feeding Wrens investigating exposed tree roots and around the base of bushes.

WATCHING TIPS

Wrens are rather discreet when not singing so it takes patience to watch them. They search log piles and dry-stone walls for insects. They may come to bird food in the winter (especially mealworms and fatty pellets) – try scattering food on the ground close to cover.

Prunella modularis — **Dunnock**

The only British member of its family, the Dunnock is a common garden bird, found throughout the UK and Ireland.

Length 14.5cm

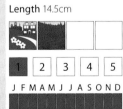

| 1 | 2 | 3 | 4 | 5 |

J F M A M J J A S O N D

How to find it

■ **Timing** Dunnocks may be seen all year round in the same habitat. Singing males are particularly obvious in spring.

■ **Habitat** This species can be found in gardens (especially those with 'wild' areas), in woodland, around hedgerows and in any other places with plenty of bushes and trees.

■ **Search tips** If you feed the birds you will probably have seen Dunnocks on the ground, picking up discarded bits from hanging feeders or foraging under overhanging bushes. Look out for males giving their simple warbled song from high and often prominent perches in spring.

WATCHING TIPS

Supply food on the ground to attract Dunnocks. Belying their rather drab appearance, these birds have very interesting and complex sociosexual behaviour, which may be easily observed from the garden. Both sexes may court and form lasting 'pair bonds' with multiple mates at the same time.

Robin

Erithacus rubecula

This much loved and attractive garden bird is easy to find all year round throughout the UK and Ireland.

Length 14cm

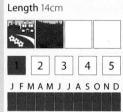

| 1 | 2 | 3 | 4 | 5 |

J F M A M J J A S O N D

How to find it

■ **Timing** Robins can be seen everywhere all year round, and they also sing throughout the year. They are most obvious in gardens in winter.

■ **Habitat** Any habitat with a good amount of trees and/or bushes can support breeding Robins, with gardens and deciduous woodland perhaps the most productive habitats.

■ **Search tips** These birds are easy to find, often perching pertly on prominent perches, and giving their distinctive and lovely song year-round and (especially in towns with street lighting) even sometimes at night. They will become extremely confiding in places where they have frequent human contact.

WATCHING TIPS

By offering high-value foods like mealworms, you will not only attract Robins to your garden but also potentially win their trust, so that they will feed from your hand. Positioning an open-fronted nest box behind a thick bush or hedge may tempt them to nest in your garden.

Bluethroat

Luscinia svecica

This lovely bird is a rare passage migrant in spring and autumn, mainly to east coasts. It has also bred on occasion.

Length 14cm

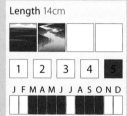

| 1 | 2 | 3 | 4 | 5 |

J F M A M J J A S O N D

How to find it

■ **Timing** Spring Bluethroats are most often found in May, while September and early October is best for autumn migrants.

■ **Habitat** These birds are mainly found near the east coast, especially on headlands, in habitats with some patchy scrubland and open areas of short grassland where they forage.

■ **Search tips** Visit east-coast headlands after easterly winds for your best chance of finding a Bluethroat. Look for a small and darkish bird with a Robin-like upright stance and hopping gait, feeding on open grassy ground or disappearing into the undergrowth.

WATCHING TIPS

Spring birds may linger for several days and (if male) sing and show signs of establishing a territory, though only occasional breeding in Britain has occurred. Autumn birds are often secretive and you may require great patience as they can lurk for hours out of sight in a bush.

Luscinia megarhynchos

Nightingale

This shy songster is a summer visitor, found south of a line from the Severn to the Humber. Migrants occasionally turn up elsewhere, but reports of singing Nightingales away from this area (or in winter) are usually cases of mistaken identity.

Length 16.5cm

| 1 | 2 | 3 | 4 | 5 |

J F M A M J J A S O N D

How to find

■ **Timing** Nightingales are generally first heard singing in mid-April, and continue to sing regularly until late May or early June. Thereafter, they become very hard to find, although dispersing juveniles in August and September are a little showier than the adults.

■ **Habitat** As long as there is an abundance of dense, low, bushy vegetation, Nightingales can be found in both woodland and more open scrubby habitats. Migrants may be found in more sparsely vegetated areas but will seek out what cover they can find.

■ **Search tips** This species is best located by its song, given by day and night, which is an unmistakable series of rich, throbbing, liquid notes. However, hearing the song is no guarantee that you will see the songster. The bird will usually be well hidden in thick scrub, and will be disinclined to move. If you get too close and flush it you may glimpse a reddish tail as it flies quickly to a new hiding place, but wait quietly and it may move to a better spot. Searching the same areas in late summer could produce clearer views of young birds.

Super sites

1. RSPB Highnam Woods
2. Little Paxton
3. RSPB Minsmere
4. RSPB North Warren
5. RSPB Northward Hill
6. RSPB Pulborough Brooks
7. RSPB Blean Woods

WATCHING TIPS

These birds are so skulking in their habits that watching them in any meaningful way is very difficult for the average birdwatcher. However, the ease with which you can enjoy their glorious song does help compensate for this. With patience and careful observation, you can become familiar with an individual male's routine and preferred singing places, and work out whether there are spots from where you can have a reasonable view. The adults are extremely discreet at the nest, but if you are lucky you may see them feeding fledged chicks in mid-summer.

Black Redstart

Phoenicurus ochruros

This distinctive small bird occurs in Britain as a rare breeding bird, mainly in urban areas in southern, central and north-western England, and as a rare passage migrant and winter visitor along the coasts of England and Wales.

Length 14.5cm

1	2	3	4	5

J F M A M J J A S O N D

How to find

■ **Timing** The situation is quite complex. Breeding birds occupy their territories between March and May, and stay until September. Passage migrants come through in March and April, and are returning between September and November, with October the peak month. Overwintering birds are present between October and March.

■ **Habitat** The Black Redstart has a preference for uneven rocky terrain, which includes sea cliffs, but may also be found in the more neglected industrial parts of cities. It will also use modern buildings provided there are scattered weedy plants around to provide a population of insects, and a complexity of nooks and crannies for nest-sites. On passage it may be seen on coastal fields or beaches, and even occasionally in gardens.

■ **Search tips** Finding Black Redstarts in cities is a matter of listening for the song, a twittered phrase followed by a bizarre gravelly, grinding rattle. Scan along rooftops and on projections like aerials for the singing bird. At many urban sites the birds nest at a considerable height, and may not be seen from street level. If searching on rocky coasts, look for places where the cliff is crumbling or there are deep fissures in the rock.

Super sites

1. Lindisfarne NNR
2. Sizewell B Nuclear Power Station
3. East London
4. Berry Head
5. Portland Bill
6. Weymouth
7. RSPB Dungeness

WATCHING TIPS

If you work in a city centre office block, you may enjoy better views of Black Redstarts than most. With more property owners exploring green initiatives like rooftop gardens, the Black Redstart population is well placed to increase, and lucky city workers could enjoy eye-level views of breeding birds. They are charming birds to watch, bold and lively with elegant fly-catching skills. On the coast, beware of confusing migrating females of the two redstart species – the differences are subtle.

Phoenicurus phoenicurus

Redstart

This beautiful bird is a summer visitor and passage migrant. It uses completely different breeding habitat to the closely related Black Redstart. It breeds in much of Scotland, Wales and England but is very rare in Ireland and Northern Ireland.

Length 14cm

1	2	3	4	5

J F M A M J J A S O N D

How to find

■ **Timing** Redstarts reoccupy their breeding habitat in April and remain there until early September. Passage migrants from elsewhere in Europe stop off along the east coast in September and October.

■ **Habitat** Mature oak woodland is the best habitat in which to look for Redstarts – both low-lying and upland woods are suitable. Other wooded habitats including mixed woodland, native Scots pine forest, tree-scattered parkland and copses and hedges in farmland can also support breeding populations. On migration they are often found along the east coast, where they are drawn to any bushy or scrubby cover.

■ **Search tips** Redstarts can be difficult to see, as they feed and sing high in the trees and the leaf canopy is well grown by the time they appear. Listen for the pleasant, prolonged and rather variable song, not readily confused with that of any of the other common woodland birds, and scan the high twigs and branches for the bird moving about. In migration periods check any bird seen in or around coastal bushes.

Super sites

1. RSPB Glenborrodale
2. RSPB Inversnaid
3. RSPB Ken-Dee Marshes
4. RSPB Haweswater
5. RSPB Coombes and Churnet
6. RSPB Carngafallt
7. RSPB Nagshead

WATCHING TIPS

Trying to watch Redstarts in the woodland canopy could lead to a cricked neck, as they spend much of their time high up. At a few nature reserves there are elevated 'tower' hides looking into the canopy, which could provide more comfortable views. It may also be worth waiting by a woodland pond for birds coming to drink. Female Redstarts and female Black Redstarts may occur together at coastal migration hotspots, when a clear view is needed to separate them (but Black Redstarts are less skulking).

Whinchat

Saxicola rubetra

This is a small and perky chat, similar to the Stonechat, which visits us in summer. Its distribution as a breeding bird is strongly biased to the north and west. It is also a widespread passage migrant, especially in autumn.

Length 12.5cm

1	2	3	4	5

J F M A M J J A S O N D

How to find

■ **Timing** Whinchats arrive on their breeding grounds in April, and depart in August or September. During September they are found on migration away from the breeding range, both inland and on the coast.

■ **Habitat** This species uses somewhat open countryside, both upland and lowland but with a preference for the latter. It needs plenty of scrub in the form of hedgerows or bushes. It likes hay meadows and fields of set-aside with a diverse growth of weedy plants, which attract the insects it feeds on. Migrating birds are attracted to similar habitat types, especially on the east coast but also well inland.

■ **Search tips** Both Whinchats and Stonechats have the helpful habit of preferring to perch on elevated, very visible points – on the top of a bush rather than within it, for example. When in suitable Whinchat habitat, look out for a small, short-tailed bird with an upright stance, posing on a wire fence, drystone wall or other prominent spot. On migration, birds are often encountered in small groups. Listen for the male's variable song, which is a jumble of sweet and scratchy notes.

WATCHING TIPS

Whinchats are quite showy birds and you can usually get quite good views of them both on the breeding grounds and on migration without too much difficulty. Be careful not to approach birds too closely in the breeding season – signs of agitation (which may include a bird feigning injury as a distraction display) are your cue to back off. Autumn Whinchats can be confused with Stonechats, as young Stonechats may show a supercilium, but it is never as prominent as that of a Whinchat. By winter there is no confusion, as the Whinchats will all have migrated.

Saxicola torquatus

Stonechat

This chat is closely related to the Whinchat, but is present in Britain all year round. It is most numerous in Scotland, Wales and western England and around the Irish coastline, and many overwinter along the east coast.

Length 12.5cm

| 1 | 2 | 3 | 4 | 5 |

J F M A M J J A S O N D

How to find

■ **Timing** Across much of Britain Stonechats are easy to find throughout the year, although in late spring to early summer when attending small chicks they become more discreet. On the east coast they become much more numerous between October and March.

■ **Habitat** The Stonechat likes open countryside with plenty of scrub and bushes, and prefers lowlands to uplands. Classic Stonechat habitat is fields with clumps of gorse bushes, or heather-dominated heathland with scattered bushes and small trees. It may be seen in gardens in quiet villages, for example on the west coast of Scotland, if there is suitable habitat in the nearby vicinity.

■ **Search tips** Stonechats draw attention both by their habit of sitting on the tops of bushes or other prominent spots, and by their distinctive 'weet-chack-chack' call (the 'chack' part sounding like two pebbles tapped together). Even seen against the light they are distinctive, with a large-headed and short-tailed silhouette. When disturbed they will usually just retreat to the next bush back. When walking on heathland, look out for Stonechats anywhere where the heather is interrupted by stands of gorse.

WATCHING TIPS

These birds are fairly unworried by people and if you sit still and quiet you should be able to watch them quite easily. From early summer you could see newly fledged young birds being fed by their parents – a second and sometimes third brood will follow. Through autumn and winter they are often seen in small loose flocks, and you can compare male and female plumages at different points in the moult cycle. Occasionally white-rumped birds of the Siberian subspecies (sometimes considered a separate species) are found in Britain, mostly in autumn.

Wheatear

Oenanthe oenanthe

A summer visitor that breeds in rocky uplands in Scotland, Wales, Ireland, Northern Ireland and northern and western England, the bold and striking Wheatear can be seen almost anywhere when on its spring or autumn migration.

Length 14.5–15.5cm

| 1 | 2 | 3 | 4 | 5 |

J F M A M J J A S O N D

How to find

■ **Timing** The first Wheatears appear in March, mainly along the south and south-east coasts. They may stay a few days before moving on to their breeding grounds. Return migration begins in August after the breeding season has ended, and again migrants may stay some time on stop-off points, especially at the south and south-east coasts.

■ **Habitat** Open grassy landscapes suit this mainly ground-feeding bird, and migrants are usually seen on well-grazed (by livestock or rabbits) or mown grass, including recreation grounds, as well as quieter areas. For breeding it needs similarly open and grassy places but with exposed rocks or drystone walls with crevices in which it can nest.

■ **Search tips** When walking on the clifftops by the seaside in early spring or mid-autumn, look out for Wheatears on any grassy patches or even around small ornamental gardens on the seafront. They run across the grass when foraging, pausing in an alert and upright posture, and will often perch on obvious elevated perches. In flight it shows a striking and distinctive black-and-white tail and rump pattern.

WATCHING TIPS

For many coast-based birdwatchers, seeing the first Wheatear of the year is a sign that spring has arrived. Spring males are particularly attractive. The particularly keen-eyed will be looking out for the Greenland Wheatear (subspecies *O. o. leucorhoa*), a larger and brighter bird that comes through slightly later than our own breeding birds. The Wheatear's showy character means that it is a relatively easy bird to watch, both on its breeding grounds and when on a migration stop-off.

Turdus torquatus

Ring Ouzel

This rather scarce thrush is a summer visitor to Britain. It breeds in the uplands in northern and western parts of England, Scotland and Wales, and around the Irish coasts, but may be seen elsewhere when on migration.

Length 23–24cm

| 1 | 2 | 3 | 4 | 5 |

J F M A M J J A S O N D

How to find

■ **Timing** Migrant Ring Ouzels arrive on the south and east coasts in March and make their way north and west to the breeding grounds. When the breeding season comes to an end in August both adults and juveniles disperse and move south-east towards the coast to begin their return migration.

■ **Habitat** Breeding habitat is remote but usually rather sheltered rocky uplands, often around crags and gullies, with scattered scrub and small trees. When on migration they may be found on both lowlands and uplands, usually feeding in fields with short turf and bushes or trees nearby, to which they can retreat if disturbed.

■ **Search tips** Compared to the more common and familiar thrushes this is a rather shy bird. Walk slowly when searching, and scan ahead often, looking along rock edges and in trees and bushes. Males choose elevated spots from which to sing; listen for the simple but rich and fluty song. When looking for migrants, concentrate on open grassy areas, scanning for birds feeding on the turf, and carefully check all distant 'Blackbirds' that you see.

Super sites

1. Beinn Eighe
2. Findhorn Valley
3. RSPB Haweswater
4. Forest of Bowland
5. Upper Derwent Valley
6. RSPB Lake Vyrnwy
7. Stanage Edge
8. RSPB Carngafallt

WATCHING TIPS

As this species is both shy and in decline as a breeding bird, particular care should be taken to avoid disturbance in the breeding season. Watch from a good distance, and try to conceal your outline. With good views the difference between Ring Ouzel and Blackbird are obvious, but beware of confusion with leucistic (showing aberrant white plumage) Blackbirds. Any apparent Ring Ouzels seen in urban settings, or anywhere in winter, should be scrutinised with particular care.

AUTUMN PASSERINE MIGRATION

Of all the birds that visit Britain for the summer, the majority are
passerines or songbirds. They arrive between March and May,
and start their return journeys south (mostly to Africa) as early as
August. You may see them in spring or autumn, but for several
reasons the autumn migration is much more pronounced and
noticeable. Looking for autumn migrants is an exciting part of
birdwatching, not least because of the chance of finding a rarity.

Waxwings

Incredible journeys

Spring migrants tend to move fast. They are experienced adult birds, eager to get
back to their breeding grounds and claim the best territories. In autumn, however,
the majority of migrants are young birds of the year, which have never migrated
before. They take longer, make more mistakes, and there are many more of them
than there will be the following spring. They are also vulnerable to the vagaries of
the weather – easterly winds will drive migrants from the near continent over to
our shores.

   The best time and place to look for autumn migrants is on an east coast
headland (or in Orkney or Shetland) in the early morning, when the wind is
coming from the east. Birds migrating south and south-east on the continent
will be pushed out over the North Sea, and will touch down on the first dry land
they reach. The result is a 'fall' – a mass arrival of migrants. In such conditions, it
can seem that every bush and tree is alive with small birds, and while most of
them will be commoner species there is likely to be the odd rarity among them.
They will feed up to recover their strength, and many will depart that night if the
weather is clear.

   Other coasts are not without interest when it comes to passerine migration.
South coast headlands are perhaps more associated with spring arrivals, but can
be 'launch pads' for southbound birds. West coast headlands and islands are most
likely to be productive when there have been westerly winds, and following very
strong westerlies these sites will usually be an arrival of a handful of extremely
rare passerines and other birds from North America. The Isles of Scilly are
particularly notable in this respect.

Inland passage

It isn't just birdwatchers on the coast that can enjoy autumn passerine migration.
Many of our summer visitors will filter southwards through the countryside,
sometimes pausing for several days at particularly good feeding sites. You may
find that your local woodland is suddenly full of Chiffchaffs for a couple of weeks
in September, or there are half a dozen Wheatears in the fields and Whinchats
along the hedgerow.

   Watching migration actually in progress is an interesting activity, known
among birdwatchers as observing visible migration. To try it yourself, check a
local map and find the highest accessible ground nearby. Position yourself on this
hilltop at dawn with binoculars and notebook at the ready. You may be surprised
by the species you see (and hear) going overhead. Some of the classic birds seen
when watching visible migration are not generally regarded as migrants, but do
demonstrate localised movement – they include Meadow Pipit, Skylark, Siskin and
Woodpigeon. However, as with watching at the coast, there is always the chance
of something unusual.

Redwing

WINTER ARRIVALS

It is not just departing migrants that you'll notice in autumn. Our winter visitors – primarily Fieldfares, Redwings and Bramblings – make up a significant part of autumn passerine migration, turning up first on the most north-easterly coastlines and gradually making their way south and west. Occasionally they will bring with them a real rarity, such as a Black-throated Thrush. Waxwings arrive in a similar pattern, although their numbers are much more variable.

Where to watch

Here are some classic coastal sites to try for autumn passerine migrants:

- Lands End
- Portland Bill
- Beachy Head
- RSPB Dungeness
- Landguard Point
- Holme
- Gibraltar Point
- Donna Nook
- Spurn Point
- Flamborough Head
- Filey Brigg
- Isle of May
- North Ronaldsay
- Fair Isle

Red-breasted Flycatcher

Blackbird
Turdus merula

This very familiar bird is extremely common both in gardens and in the wider countryside throughout Britain.

Length 24–25cm

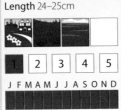

1	2	3	4	5

J	F	M	A	M	J	J	A	S	O	N	D

How to find it

■ **Timing** Blackbirds can be seen year-round, although numbers more than double in winter as migrants arrive from the continent.

■ **Habitat** Any environment with a mix of open ground and cover in the form of trees and bushes will hold Blackbirds. They forage particularly on grassland with short turf on soft ground, and in autumn and winter in and around fruiting trees and bushes.

■ **Search tips** Even small urban gardens will attract Blackbirds, especially if there is some lawn and some berry bushes or fruit trees. They are also easy to find in the countryside, especially singing males in spring.

WATCHING TIPS

Garden Blackbirds are interesting to watch year-round. Both sexes fight for territory in spring, and the successful pair may well nest in a garden bush, rearing two or three broods. In winter you may see small groups feeding on the lawn or taking windfall fruit.

Song Thrush
Turdus philomelos

The smaller of our two resident spotted thrushes, the Song Thrush is common throughout Britain, but has declined recently.

Length 23cm

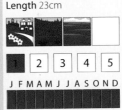

1	2	3	4	5

J	F	M	A	M	J	J	A	S	O	N	D

How to find it

■ **Timing** Song Thrushes are here all year round, although are more likely to visit gardens in autumn and winter.

■ **Habitat** Any habitat that supports Blackbirds will also be good for Song Thrushes. Gardens, woodland and open countryside with hedgerows and scattered small copses are all attractive to them.

■ **Search tips** Look for Song Thrushes feeding on your lawn or other short turf in town and country. The beautiful mellow song of repeated phrases is usually given from a tree. They also forage on ploughed fields, and along hedgerows and in orchards.

WATCHING TIPS

These thrushes are rather shy but can be watched easily in the garden. Never use slug pellets, to avoid poisoning Song Thrushes. Look out for them breaking open snails by swinging them against a stone. Putting out soft apples and pears will attract them, especially in a cold snap.

Turdus pilaris — **Fieldfare**

This large and colourful thrush is a winter visitor to Britain, potentially found everywhere but more numerous in the north and east, especially earlier in the winter season. It is also an extremely rare breeding bird in Scotland.

Length 25.5cm

1	2	3	4	5

J F M A M J J A S O N D

How to find

■ **Timing** The first Fieldfares arrive in early October or even the end of September, with large flocks sometimes making landfall along the east coast. By mid-winter they will have spread well inland. The last birds to leave in spring hang on well into April.

■ **Habitat** Fieldfare flocks seek out berries and other fruit in autumn and early winter, descending upon hedgerows and also visiting orchards and gardens. As winter progresses and berry supplies run low, they often move on to ploughed or grazed farm fields to forage for soil invertebrates, and lone birds may take up long-term residence in gardens.

■ **Search tips** This is a gregarious and vocal species so usually quite easy to find. Listen for the chuckling or clucking calls of a flock on the move. When out in the countryside, check along hedgerows and on the ground around fruit trees, and also watch out for flocks passing overhead, when the solid black tail and clean white underwings and lower belly is noticeable and distinguishes them from the similar-sized Mistle Thrush.

WATCHING TIPS

Fieldfares are shy birds and quick to take alarm. The exception comes in very cold winter weather when desperately hungry birds may tolerate close approach while they are feeding, but this should of course be avoided. Placing food such as cut-up soft apples and pears and soaked dried fruit on the ground in your garden is a good way to attract them. Sometimes a lone Fieldfare defends a garden as a feeding territory. Any Fieldfares seen between mid-May and August are potentially breeding locally, so the sighting should be reported to the local recorder.

Redwing

Turdus iliacus

This thrush is a common winter visitor, first arriving in the north and east and remaining most numerous there, but spreading to the whole of Britain as autumn progresses. It is also a very rare breeder in the Scottish Highlands.

Length 21cm

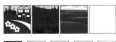

| 1 | 2 | 3 | 4 | 5 |

J F M A M J J A S O N D

How to find

■ **Timing** Along the east coast the first Redwings are seen in September, but the main arrival across the wider countryside is in October. Numbers begin to thin out in March and nearly all will have moved on by the end of April.

■ **Habitat** Like Fieldfares, Redwings roam the countryside in search of food, in the form of fruit and invertebrates, and they will often flock with Fieldfares. Scrub, woodland edges and orchards attract them in autumn and early winter especially, and they spend more time foraging on farm fields in late winter and spring. They will also visit gardens if there are berry-bearing shrubs like cotoneaster and pyracantha.

■ **Search tips** Listen in October by night and day for migrating Redwings flying overhead – their thin and rather harsh 'tzee' flight call is very distinctive. The outline is rather compact and short-tailed for a thrush, and could be mistaken in silhouette for a Starling. Search hedgerows and berry-bearing trees in towns and gardens, and scan fields, especially along the margins near hedges.

WATCHING TIPS

A garden well stocked with fruiting bushes stands a good chance of attracting Redwings in winter. Like Fieldfares they are shy and wary birds but will come to towns and gardens more when there is a snow fall or significant freeze. It is quite common to see mixed flocks of Redwings, Fieldfares and sometimes other thrushes. By early spring many of the large flocks will have broken up as concentrated food supplies become more difficult to find, and before the return migration you may even hear males singing in April, a simpler and less rich song than our resident thrushes.

Turdus viscivorus

Mistle Thrush

This is our largest species of thrush, and a common resident species throughout Britain except on some of the northernmost island groups. It is often confused with Song Thrush but with a little practice is easy to identify.

Length 27cm

1	2	3	4	5

J F M A M J J A S O N D

How to find

■ **Timing** Mistle Thrushes can be seen all year round, although are at their least obvious in late spring and early summer. Post-breeding flocks form in late summer to early autumn, but in winter it may be solitary and territorial.

■ **Habitat** This thrush feeds on open ground and in bushes, but needs tall trees for nesting. Ideal habitat combines all of these features – it can be common in parkland, suburbia, forest edges, and open farmland with hedgerows and small areas of woodland. It is often seen searching for prey on playing fields and village greens.

■ **Search tips** Not as shy and skulking as most other thrushes, the Mistle Thrush can often be found feeding on the ground well out in the open. It is fiercely territorial and frequently gives itself away with a loud rattling call when seeing off a rival or mobbing a predator. Singing males choose quite exposed and elevated spots, and are noted for continuing to sit out and give their powerful song in weather bad enough to deter other thrushes from singing.

WATCHING TIPS

These bold thrushes are easy to see, with birds in town parks usually quite approachable. They are less frequent visitors to gardens than Song Thrushes or Blackbirds but if your garden has tall trees and a ready food supply you may attract them. Their habit of establishing winter feeding territories is unusual among songbirds. A single bird or a pair will vigorously defend a food source – for example a berry-covered holly tree – against other birds (usually other thrushes), only conceding defeat if overwhelming numbers of intruders arrive.

Cetti's Warbler

Cettia cetti

This shy marshland warbler first bred in Kent in the early 1970s, but has since spread from the original site along the southern and eastern coasts of England and to south Wales, although successive cold winters could curtail its spread.

Length 13.5cm

| 1 | 2 | 3 | 4 | 5 |

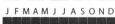

J F M A M J J A S O N D

How to find

■ **Timing** You can see (or more likely hear) Cetti's Warblers all year round. They may abandon more inland areas when there is a severe freeze.

■ **Habitat** This is a bird of wet, lushly vegetated lowland habitats – reedbeds, marshes, riversides and fens. Anywhere near sheltered water (fresh, brackish or salty) that has plenty of dense cover is potentially good habitat for them, and sites near the coast are especially good as they are less likely to be affected by severe weather.

■ **Search tips** Cetti's Warblers sing all year round, giving a series of short distinctive rich, loud and explosive notes. On a typical visit to a good Cetti's site you will hear several birds singing, but to catch a glimpse of one of them requires great patience and a bit of luck. The species is extremely skulking, keeping deep within thick tangled vegetation and making short, low flights between bushes. Often all that is seen is the broad red-brown tail as the bird dives into cover. Try watching from hides overlooking water margins, especially in winter – look out for a dark brown bird feeding discreetly on or near the ground.

Super sites

1. RSPB Strumpshaw Fen
2. RSPB Minsmere
3. RSPB Marazion Marsh
4. RSPB Radipole Lake
5. RSPB Lodmoor
6. RSPB Brading Marshes
7. Oare Marshes
8. Stodmarsh NNR
9. RSPB Dungeness

WATCHING TIPS

Just seeing this bird is a real challenge, so to watch it at any length is extremely difficult. However, if you have Cetti's Warblers on your local patch, putting in some time to become familiar with their habits could reap rewards. Males typically sing from several preferred perches within a small area in rotation, slipping very discreetly between them. Once you have worked out which spots are used, you may be able to find viewpoints from which to see the singing bird. As a resident species, it suffers heavy losses in hard winters and is liable to suddenly disappear from previously occupied sites.

Grasshopper Warbler

This is a very shy and skulking warbler, currently suffering a period of severe decline in Britain. It is still quite widespread in summer in England, Wales, the southern half of Scotland and most parts of Ireland and Northern Ireland.

Length 12.5–13.5cm

| 1 | 2 | 3 | 4 | 5 |

J F M A M J J A S O N D

How to find

■ **Timing** Grasshopper Warblers are usually first noticed when they settle on their territories and the males begin to sing. May and June are the best times to look for singing birds, while in September passage migrants turn up on the coastline but are skulking and difficult to see.

■ **Habitat** Though often associated with watery habitats the Grasshopper Warbler is not really a bird of reedbeds and marshland, but the slightly drier scrub or fen that borders the marshes, and sometimes drier habitats like heathland, rough weedy grassland or the edges of pine plantations. It requires plenty of dense low-level vegetation.

■ **Search tips** Listen for the song in spring and early summer. The thin, dry, continuous sound resembles that of a turning angler's reel, and it is most likely to be heard early in the morning or at dusk (and also at night). The singing bird may be sitting in full view, although its dull streaky plumage makes it difficult to spot. If you hear the song, stand back and scan all bushes and tall plants nearby – the bird can 'throw' its voice so working out where the song is coming from is not straightforward.

WATCHING TIPS

If you're lucky, when you locate a singing Grasshopper Warbler it will be on view (although they will also sing from well within thick cover), quite near the ground, and will stay put for a while if you don't try to get too close. If it does move, you may be able to watch it quietly moving in the vegetation, creeping about like a mouse. Grasshopper Warblers on migration will keep in cover but if you stay still and quiet you may be able to watch them feeding on the ground or occasionally fly-catching.

Savi's Warbler

Locustella luscinioides

Closely related to the Grasshopper Warbler, Savi's Warbler is a very rare breeding bird and passage migrant to Britain.

Length 14cm

| 1 | 2 | 3 | 4 | **5** |

J F M A M J J A S O N D

How to find it

■ **Timing** Most Savi's Warblers are found in spring or early summer, when the males are singing.

■ **Habitat** More tied to wetland habitats than the Grasshopper Warbler, this species is almost invariably found in reedbeds around fresh water.

■ **Search tips** This warbler has a 'reeling' song, rather like the Grasshopper Warbler but lower pitched, more powerful and given in shorter bursts. Listen to recordings to get familiar with the two songs.

■ **Super sites 1.** Hickling Broad, **2.** RSPB Radipole Lake and RSPB Lodmoor, **3.** Stodmarsh NNR.

WATCHING TIPS

This warbler formerly bred regularly in south-east England in small numbers, but now is rather sporadic, so any signs of breeding behaviour should be reported. It looks rather similar to Reed Warbler but good views reveal many subtle differences.

Aquatic Warbler

Acrocephalus paludicola

This warbler is one of the most endangered species in Europe. It is a very rare passage migrant to southern Britain.

Length 13cm

| 1 | 2 | 3 | 4 | **5** |

J F M A M J J A S O N D

How to find it

■ **Timing** Most Aquatic Warblers in Britain are found in summer, with late August the best time. They are more likely to venture into view on days that are not too windy.

■ **Habitat** Sightings have mainly come from areas with reedbeds and freshwater marshland.

■ **Search tips** This is a shy and skulking warbler, and must often be overlooked (most records have been birds trapped by ringers). Watch from a hide overlooking reedbeds. Look out for it moving quietly among the reeds, and beware confusion with juvenile Sedge Warblers, which will be present at the same sites.

■ **Super sites 1.** RSPB Marazion Marsh, **2.** RSPB Radipole Lake and RSPB Lodmoor.

WATCHING TIPS

Because of its global rarity, its migratory staging posts in Britain are very important from a conservation point of view, and sightings should always be reported. If possible take photographs, to confirm that you have found an Aquatic and not a Sedge Warbler.

Acrocephalus schoenobaenus

Sedge Warbler

A common summer visitor to almost the whole of the UK and Ireland, the Sedge Warbler is a lively small bird of marshy wetlands. Its song is similar to that of the Reed Warbler but with practice easy to recognise.

Length 12–13cm

| 1 | 2 | 3 | 4 | 5 |

J F M A M J J A S O N D

How to find

■ **Timing** Sedge Warblers are probably easiest to find in May and early June when the singing of the males is at its peak. They are also much in evidence in August when reedbeds can seem full of juvenile birds.

■ **Habitat** Look for Sedge Warblers in reedbeds and also in drier but still lushly vegetated areas on the edges of marshland. On migration, they may be found on drier habitat near the coast.

■ **Search tips** Listening for the male's song is the easiest way to find this bird. The song is more excitable and varied than that of Reed Warbler, and more likely to be heard from vegetation near the reeds than actually within the reedbed. The song perch is often on the edge of a small bush and may be in full view. If you cannot see the bird, wait for a while and look out for a bird taking flight and rising steeply before dropping back down – males make frequent songflights in between singing from perches. In autumn, look in the reedbeds, as the birds feed very heavily on reed-dwelling aphids before they migrate.

WATCHING TIPS

Sedge Warblers are usually fairly confiding and approachable. Singing males will generally stick to the same spot for several minutes, often twirling around a vertical twig or stem, and will be unworried by your presence provided you keep still and quiet. The songflight is used to move to a new perch. Experience is needed to learn the difference between Sedge and Reed Warbler song but Sedge song is more variable. If the song includes many dry clicking or scratching sounds, it is probably a Sedge Warbler. Numbers of juvenile Sedge Warblers in late summer can be incredibly high in large reedbeds.

Marsh Warbler
Acrocephalus palustris

This warbler is a very rare breeding bird and passage migrant in Britain, with most pairs in the south-east of England and around the Severn Valley. It is extremely similar to the Reed Warbler, and most easily identified by its song.

Length 13cm

1	2	3	4	5

J F M A M J J A S O N D

How to find

■ **Timing** Marsh Warblers are among the latest arriving of all our summer visitors, with some not arriving until early June. On windy days they will keep well down in the vegetation, so are best looked for in still weather.

■ **Habitat** This bird breeds in marshland with reedbeds and other tall, thick marginal vegetation.

■ **Search tips** As with the other *Acrocephalus* warblers, hearing the song is key to finding this species. The Marsh Warbler is a noted songster, producing a much more melodious and varied song than the Reed Warbler, although just to complicate things Reed Warbler song differs from bird to bird. There are also slight differences in appearance. Finding the singing bird is a matter of very careful scanning and searching through the vegetation. Use binoculars and try to look through the vegetation – altering your viewpoint very slightly could make the difference between seeing the bird and it being totally obscured. Also look out for small movements of the vegetation to pinpoint the bird. Marsh Warblers do occur on migration, but their identification is even more difficult than with spring birds..

Super sites

1. Shetland
2. Ornkey
3. Flamborough Head
4. Spurn Point

WATCHING TIPS

Marsh Warblers are difficult to see well, but the song is a real treat to enjoy. Its most interesting feature is the large amount of mimicry incorporated. If you know your bird songs and calls you will probably recognise many species, but there will also be a proportion of mimicked songs and calls from African birds, species that the Marsh Warbler encounters on its wintering grounds. If you see a non-singing possible Marsh Warbler, pay particular attention to its general plumage tones, the size of its bill, long primary projection, and the colour of its legs. Migrants are regularly found on the Scottish islands.

Acrocephalus scirpaceus

Reed Warbler

This rather plain and shy reedbed and marshland warbler is a common and widespread summer visitor in England – especially the south and east – and Wales. It is much scarcer and more localised in Scotland, Ireland and Northern Ireland.

Length 13cm

| 1 | 2 | 3 | 4 | 5 |

J F M A M J J A S O N D

How to find

■ **Timing** Reed Warblers will be on territory and singing from mid-April. They continue to sing well into summer and even early autumn. They are easier to see on still, fine days.

■ **Habitat** Any reedbed in the right parts of Britain, whether large or very small, can attract Reed Warblers. Large reedbeds can hold quite dense populations. Passage migrants in spring will make use of very small areas of reeds or sedges, even occasionally in town parks, and males may sing and hold territory briefly before moving on to more suitable habitat.

■ **Search tips** This warbler nearly always sings from a quite well-hidden spot in the reeds. Listen for the rather monotonous song and then scan the reeds, looking out for movement. Unless disturbed, a singing bird will keep to the same spot for long spells, so try different viewpoints, ideally using some sort of cover as the warbler will drop out of sight if it feels threatened. Newly fledged youngsters, seen in June and July, will sometimes sit in view as they wait for food deliveries from their parents.

WATCHING TIPS

Reed Warblers are easiest to watch from a hide that gives a slightly elevated view over the reeds, making the birds easier to find. Failing that, try watching through a gap in bushes bordering a reedbed. The birds stay low on windy days but on still days may be watched climbing among the reeds, gathering fronds of seed heads for nesting material, and perhaps even tending their nests, which are woven between several reed stems. Look out for Cuckoos in an area with many Reed Warblers, as Reed Warbler is one of the main Cuckoo hosts.

Blackcap

Sylvia atricapilla

An attractive woodland warbler, the Blackcap is a common and widespread summer visitor, and a more localised winter visitor (two separate populations are involved). It is one of the few British warblers in which the sexes look quite different.

Length 13cm

1	2	3	4	5

J F M A M J J A S O N D

How to find

■ **Timing** The Blackcaps that visit us for spring and summer establish their territories in mid-April, and are present until October when they migrate to Spain and Africa. They are easiest to see in spring. Meanwhile, part of the central European breeding population migrates to Britain for winter – these birds are most readily seen in mid-winter.

■ **Habitat** Deciduous woodland is Blackcap breeding habitat, with tall trees for song perches and a low scrubby understorey for nesting sites. Birds preparing to migrate in autumn visit berry-laden hedgerows. Wintering birds frequently visit gardens, where they are attracted to fat-based foods offered on bird tables and in feeders.

■ **Search tips** The lovely rich, fluting song gives away the presence of a male Blackcap. Look for the singing bird in treetops (much easier when the birds have first arrived and the trees have not yet come into full leaf) and bushes. In autumn, any berry-bearing shrubs are worth checking. In winter, keep an eye on your bird feeders (especially if you put out fat blocks or mealworms) for visiting Blackcaps.

WATCHING TIPS

Many people's first experience of Blackcaps is when one visits the garden feeders. Most gardens will only have one bird or a pair, as they are quite intolerant of each other – they are also very aggressive towards other garden birds and may almost monopolise a favourite feeder for weeks on end. Providing an extra feeder will help the other garden birds to feed in peace. Watching Blackcaps in spring can be hard work, as the singing birds sometimes keep deep within cover, and the rest of the time they are flighty, moving constantly from tree to tree.

Sylvia borin

Garden Warbler

A very nondescript little bird, the Garden Warbler is a common summer visitor to almost all of England and Wales, but is more localised in northern and eastern Scotland, and restricted to central parts of Ireland and Northern Ireland.

Length 14cm

1	2	3	4	5

J F M A M J J A S O N D

How to find

■ **Timing** Garden Warblers arrive on their territories in late April or early May, a little later than Blackcaps (the main confusion species when it comes to song). They are easiest to find in May when the males are singing regularly, and in late summer when feeding up before migration.

■ **Habitat** This bird uses similar habitat to the Blackcap, but especially likes quite thick woodland, such as coppices, and scrubby woodland edges, and is less commonly found in mature woodland with little in the way of understorey. It feeds on hedgerow berries in autumn. Despite the name, it is seldom found in gardens.

■ **Search tips** The first step to finding this bird is to listen for the song, and the second is to make sure that it is a Garden Warbler and not a Blackcap you are hearing. In general the Garden Warbler's song is more hurried, monotonous and continuous than that of a Blackcap, but variations in subsong can make identification difficult. It often sings from a high perch and moves restlessly around the treetop while singing. Check berry bushes in quiet hedgerows in late summer and early autumn.

WATCHING TIPS

Although lacking striking plumage features, the Garden Warbler is an attractive bird in its own way, and well worth a closer look. It may be confused with Reed Warbler or female Whitethroat, but a clear view will reveal the very beady-looking dark eyes in a very plain face, the rather thick and stubby bill, and a subtle ash-grey wash on the neck-sides. In late summer this species becomes an eating machine, converting a sizeable proportion of its body weight to fat to fuel the migratory journey. It bears a superficial resemblance to juveniles of the scarce Barred Warbler.

Barred Warbler

Sylvia nisoria

A large and stocky warbler, this species occurs in Britain as a scarce passage migrant in autumn, with most records being on the east coast and northern Scottish islands, a few from other coasts. Inland sightings are very rare.

Length 15.5–17cm

| 1 | 2 | 3 | 4 | 5 |

J F M A M J J A S O N D

How to find

■ **Timing** Barred Warblers are seen between August and November, the earliest sightings usually in the far north, with late August to late September being the peak time. Sightings are most likely during or immediately after a period of moderate or strong easterly winds.

■ **Habitat** Most Barred Warblers are found in thick coastal scrubland, where they feed on berries and whatever insects they can find. Where they make landfall in areas with little suitable habitat, they will use whatever cover they can find and may visit coastal gardens.

■ **Search tips** These birds are quite shy and skulking, but their size and weight does mean they cause a bit more disturbance among the vegetation as they feed than smaller birds do. Whenever you are on the east coast in autumn, especially on a headland, check hedgerows and bushes for this species, looking for a relatively hefty and rather pale and plain warbler with a long tail and quite a heavy bill. You may have to wait some time for a clear view but the bird will probably move to the edge of the bush from time to time.

Super sites

1. Shetland
2. Orkney
3. Flamborough Head
4. Spurn Point
5. North Norfolk coast

WATCHING TIPS

This shy warbler is a spectacular-looking bird in adult plumage, but the majority of records in the UK are dull-plumaged juvenile/first-winter birds, which look rather like oversized Garden Warblers but not quite as nondescript. Look out for the pale tips to the wing-coverts, creating a faint double wing-bar, and dark crescent-shaped markings on the undertail. The general plumage coloration is also more grey compared to the browner Garden Warbler. The best views are generally of very recently arrived birds in areas where there is not too much scrub around in which they can lose themselves.

Sylvia curruca

Lesser Whitethroat

This skulking warbler, smaller than the Whitethroat, is a summer visitor mainly to England and Wales. It is scarce in central northern England, and just reaches south-east Scotland. In Ireland it is very rare and only occurs in the south-east.

Length 13cm

1	2	3	4	5

J F M A M J J A S O N D

How to find

■ **Timing** Singing Lesser Whitethroats can be heard on their territories from late April, and by mid-June will be singing much less. Juveniles in late summer are not as shy as adults.

■ **Habitat** Look for this bird in scrubby habitats, the edges of low-lying deciduous woodland, thick tangly copses and mature hedgerows in farmland, and on lowland heath with plenty of bushes. It may also use the edges of conifer plantations on low ground. It can be found feeding on hedgerow berries in late summer and early autumn.

■ **Search tips** This bird is usually located by its song, a combination of a Whitethroat-like scratchy warble and a series of repeated buzzing 'dzee' notes, like the beginning of a Yellowhammer's song. Unfortunately it sings from a well-hidden perch deep within a bush and can be extremely difficult to spot – even when moving between bushes it is quick and discreet. You may find it easier to see in late summer, when it will be feeding up before migration and may be less skulking.

WATCHING TIPS

Separating this species from Whitethroat can be difficult. If you see a whitethroat species but cannot get a clear view, there is a good chance that it is this species, as Whitethroats will usually oblige by sitting on an open perch sooner or later. Pay particular attention to the colour of the wing-coverts, which are dull brown in Lesser Whitethroat but bright rufous in Whitethroat in all plumages, and leg colour (dark grey in Lesser Whitethroat, orange-pink in Whitethroat). Surprisingly, migrating Lesser Whitethroats occasionally visit garden feeding stations, giving great views.

Whitethroat

Sylvia communis

This sprightly and noisy warbler is a prominent inhabitant of open scrubby areas in spring and summer. It is a common breeding bird throughout Britain, apart from the high uplands of Scotland and most northerly Scottish islands.

Length 14cm

1	2	3	4	5

J F M A M J J A S O N D

How to find

■ **Timing** Whitethroats are on territory and singing from mid-April. The frequency of singing tails off through June, although the species remains conspicuous through summer as the young birds fledge.

■ **Habitat** This bird uses more open countryside than the Lesser Whitethroat. Anywhere with patches of bushes or even just tall weedy vegetation will support it. Heathlands, downlands, coastal scrub and farmland hedgerows are all good places to look. It feeds up on berries in late summer, and may be seen with other *Sylvia* warblers working its way along a well-stocked hedgerow.

■ **Search tips** Male Whitethroats often sing from exposed perches – perhaps the topmost loop of a bramble clump or a tall stem of cow parsley. Listen for the song, a somewhat scratchy and toneless warble, and look for a slim and long-tailed bird. It also performs a songflight, which may be between bushes or vertical with a return to the same perch. Females and juveniles are not especially skulking either and may be seen perched in full view – often their rather nasal scolding calls will draw your attention.

WATCHING TIPS

Whitethroats are quite bold and fairly approachable, meaning that it can be easy to enjoy good views of them going about their business, including collecting food for nestlings, or feeding newly fledged chicks. The frequent harsh, churring calls give away their activity. Although easy to find they are often very active. With good views identification is straightforward. The most useful feature to look for is the bright rufous edges to the wing feathers, which distinguish them from all other brown and grey warbler species likely to be encountered in Britain.

Sylvia undata

Dartford Warbler

A very small but long-tailed warbler, this species is an uncommon breeding bird in southern and eastern England. As a resident species, it can suffer high mortality in cold winters, but equally will increase after a series of mild winters.

Length 13cm

1	2	**3**	4	5

J F M A M J J A S O N D

How to find

■ **Timing** Dartford Warblers are on their breeding grounds all year round. In autumn and winter, dispersing young birds may turn up away from the usual breeding range, and given a good breeding season and a subsequent mild winter can colonise new areas quickly.

■ **Habitat** Lowland heath is the only habitat to support breeding populations of Dartford Warblers in Britain. The birds favour areas with extensive heather cover, and scattered gorse bushes that provide song perches for the males. The further south-west the heaths are, the more likely they are to hold onto their Dartford Warblers through severe winter weather.

■ **Search tips** When walking in heathland, keep looking around for small, dark, long-tailed birds flitting low over the heath. They will often pause on prominent perches, in a characteristic stance with the long tapered tail cocked, but are wary and quick to fly if unsettled. The song, given from an open perch, is a distinctive very dry and scratchy warble that could be confused with that of Whitethroat, but that species will generally stick to scrubbier, bushier areas rather than the open heath.

Super sites

1. RSPB Minsmere
2. Westleton Heath
3. RSPB Aylesbeare Common
4. RSPB Arne
5. New Forest
6. Thursley Common
7. Ashdown Forest

WATCHING TIPS

At the edge of its breeding range, the Dartford Warbler has a rather tenuous foothold in Britain. Its numbers fluctuate dramatically in response to winter weather conditions, meaning that it could easily disappear from even well-established sites. Birdwatchers have an important role to play in keeping track of its numbers, both at core sites and at newly colonised areas. Heathland is a scarce habitat type in Britain, but heathland restoration projects could aid this bird's spread and stabilise its numbers. If you live near such a project, this is a key species to look out for.

Pallas's Warbler
Phylloscopus proregulus

This tiny and colourful warbler from Siberia is a very rare autumn passage migrant, mainly found along the east coast.

Length 9–10cm

| 1 | 2 | 3 | 4 | 5 |

J F M A M J J A S O N D

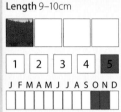

How to find it

- **Timing** The arrival of Pallas's Warblers typically coincides with easterly winds, and most are found in late October and early November. Numbers vary greatly from year to year.
- **Habitat** Most sightings are in areas of woodland on the coast. Even a few isolated trees – particularly sycamores – may be sufficient to shelter a newly arrived migrant.
- **Search tips** Look in trees on the coast after suitable weather conditions, checking every small bird you see. Pallas's Warblers are often spotted when they 'hover-feed' Goldcrest-style around the outside leaves of trees.
- **Super sites 1.** Lindisfarne, **2.** Flamborough Head, **3.** Spurn Point, **4.** Margate.

WATCHING TIPS

This warbler is beautifully and boldly patterned in green and yellow, and is much coveted by birdwatchers. It is quite confiding and may give good views, although it is also frantically active and can be hard to follow as it moves through the twigs.

Yellow-browed Warbler *Phylloscopus inornatus*

Another very rare autumn visitor from Siberia, the Yellow-browed Warbler visits the east coast in small numbers, occasionally overwinter.

Length 9–11cm

| 1 | 2 | 3 | 4 | 5 |

J F M A M J J A S O N D

How to find it

- **Timing** With its peak arrival slightly earlier than that of Pallas's Warbler, this species is most often found in October, after easterly winds.
- **Habitat** On arrival it seeks out bushes or trees, especially sycamores, and is often in company with other small birds such as Willow Warblers and Chiffchaffs. It is very scarce away from the coast.
- **Search tips** Check all small birds seen in suitable habitat in autumn, looking for a small, strongly patterned Chiffchaff-like bird with a prominent wing-bar or two (the top bar is not always obvious). Listen for its loud and frequent 'tsooeet' call.
- **Super sites 1.** Lindisfarne, **2.** Flamborough Head, **3.** Spurn Point and Donna Nook, **4.** Margate, **5.** Isles of Scilly.

WATCHING TIPS

These very active little birds can be hard work to watch, as they move restlessly through a tree or bush, constantly feeding, but be patient and you should eventually enjoy great and possibly close views as a bird in full feeding mode will be almost oblivious to your presence.

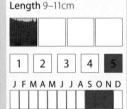

Phylloscopus sibilatrix

Wood Warbler

This is the largest and the scarcest of our three breeding *Phylloscopus* warblers. It is a summer visitor to Scotland, Wales and northern and western parts of England, and is very scarce in Ireland and Northern Ireland.

Length 11–12.5cm

1	2	3	4	5

J F M A M J J A S O N D

How to find

■ **Timing** From mid-spring to early summer, Wood Warblers will be on territory and singing frequently. They quieten down as the breeding season progresses. Passage migrants are seen in early April and in August and September.

■ **Habitat** Wood Warblers breed in mature deciduous woodland, especially oak and beech, with a continuous canopy and little in the way of understorey. They are particularly numerous in upland oakwoods. On passage in spring and autumn, they may be found in more varied places including open, scrubby habitat, both inland and on the coast.

■ **Search tips** Listening for the distinctive song is the way to find a Wood Warbler in breeding areas. There are two distinct song phases – a series of pure, descending 'teu' notes, and a rapid silvery trill, likened to the spinning of a coin. On hearing the song, scan nearby trees, checking low branches as well as the canopy. Passage migrants, which may turn up almost anywhere, may give themselves away with a penetrating 'tiu' call, lacking the upwards inflection of typical Willow Warbler and Chiffchaff contact calls.

Super sites

1. RSPB Glenborrodale
2. RSPB Inversnaid
3. RSPB Wood of Cree
4. RSPB Mawddach Woodlands
5. RSPB Carngafallt
6. RSPB Gwenffrwd-Dinas
7. RSPB Nagshead

WATCHING TIPS

This very attractive warbler is often confused with the Willow Warbler, but a clear view will reveal that it is much more brightly coloured, and bulkier with long wings and a 'top heavy' look. The nature of the breeding habitat means that Wood Warblers can be quite easy to watch as there is no thick vegetation in which they can disappear, although they are very active and restless. Watch out for them hover-feeding around the outer edges of trees, picking tiny insects from the leaves. Passage migrants are similarly hyperactive, and seek out trees or bushes in which to feed.

Chiffchaff

Phylloscopus collybita

A common warbler of lowland wood and scrub, the Chiffchaff is primarily a summer visitor to Britain but also overwinters in small numbers, mainly in the south. It is most common in southern England, and absent from parts of north Scotland.

Length 11cm

| 1 | 2 | 3 | 4 | 5 |

J F M A M J J A S O N D

How to find

■ **Timing** Chiffchaffs arrive early in spring, and can often be heard singing from early in March – they continue to sing through to September and even on autumn migration, but singing peaks around the start of the breeding season. The main autumn migration takes place through September.

■ **Habitat** These warblers are quite adaptable and will breed in a variety of habitats provided there are some trees and bushes around. Ideal habitat has a mixture of tall trees and scrub. They may be found in all kinds of deciduous woodland and in parkland and gardens. Overwintering birds often favour areas such as sheltered rivers and sewage farms where there is a sufficient population of insects throughout the winter.

■ **Search tips** The distinctive monotonous song should lead you to the first Chiffchaff of the year in early spring. The bird will probably be singing from a nearby treetop, moving constantly from twig to twig. In autumn listen for the 'hweet' contact call and a down-slurred variant – in some areas huge numbers of migrants pass through and trees and bushes may be alive with them.

WATCHING TIPS

Chiffchaffs are readily identified by song, but silent birds are easily confused with Willow Warblers. Studying both in spring, when identification can easily be confirmed, will help you to identify the non-singing autumn birds more easily. Photographing problematic individuals can help too, revealing features that aren't obvious on a fast-moving bird. Some of our wintering Chiffchaffs are not British breeding birds but belong to other subspecies from Scandinavia or even Siberia – identifying all three different forms is a challenging way to spend a winter day.

Willow Warbler

Present in spring and summer across the whole of the UK and Ireland, the Willow Warbler is perhaps the most common of all our summer-visiting species. Although found everywhere it is much more abundant in the north than the south.

Length 11cm

1	2	3	4	5

J F M A M J J A S O N D

How to find

■ **Timing** Willow Warblers arrive in late March or early April. They can be heard singing through April and May, the singing petering out (but not completely) through June. By August juveniles will be dispersing far and wide, prior to migration.

■ **Habitat** This bird will breed in all kinds of wooded habitat, from continuous mature forest to scrubland with a few scattered trees, and will breed in parks and gardens. It is more common in drier and more upland areas than the Chiffchaff but the habitat use of the two species does overlap.

■ **Search tips** As with most of the warblers, this is not a particularly showy species and is best located by listening for the song. This is a series of very sweet descending notes, usually given from a high vantage point (which may be a tree or a bush, depending on the habitat). Young birds in mid to late summer are less shy and active than their parents, and draw the eye with their very striking rich yellow underparts, often leading to misidentification as Wood Warblers (the latter species, though, always shows a clean white belly).

WATCHING TIPS

Like the other *Phylloscopus* warblers, the Willow Warbler is a very active little bird and you will need to be patient to watch it properly. However, in some of the northern parts of its range it is extremely common, which considerably increases the chances of good views. It feeds at all vegetation heights, and birds seen behaving discreetly at or near ground level may be attending a nest. Although mainly an insect feeder, it may be seen taking berries in late summer, before migration. Confusion with Chiffchaffs is common if you don't hear the song, but face pattern, wing length and leg colour all help with identification.

Goldcrest

Regulus regulus

The Goldcrest and the Firecrest are Britain's smallest birds. The Goldcrest is found year-round throughout the UK (except the northernmost Scottish islands) and Ireland, and our resident population is augmented by variable numbers of migrants in autumn.

Length 9cm

| 1 | 2 | 3 | 4 | 5 |

J F M A M J J A S O N D

How to find

■ **Timing** Goldcrests are present year-round, but numbers can be low at the start of spring if the preceding winter was very harsh. Easterly winds in autumn may bring large numbers of migrants to the east coast.

■ **Habitat** These birds use most kinds of wooded habitat including gardens, but have a particular attachment to conifers. Pine plantations, often rather devoid of bird life, may hold numerous Goldcrests, and it is common to find them in isolated pines among other trees.

■ **Search tips** Male Goldcrests sing throughout spring and summer, while winter flocks call constantly to each other as they feed. Hearing these thin and very high-pitched sounds will show you where to search for the birds, although they can still be difficult to spot among dense foliage. However, they are so active that they will almost inevitably come into view sooner or later. Some people find they are unable to hear the calls, which are close to the upper frequency range of the human hearing threshold. If this is you, try scanning around conifer trees, looking for very small, restless birds.

WATCHING TIPS

Goldcrests are single-minded in their pursuit of food, and can give astonishingly close views as they usually show no interest in humans whatsoever. Following their constant movements can still be a challenge, though. They are able to survive the winter because they can pick minute scale insects from in between pine needles – a food resource that is out of reach to larger birds – but can suffer huge mortality in a freeze. At this time they may visit garden feeders, especially if you offer fat blocks or pellets. If your garden has even just one conifer you may well have resident Goldcrests. You may be able to attract them closer by 'pishing' – making squeaky sounds with pursed lips.

Regulus ignicapillus

Firecrest

This exquisite bird, closely related to the Goldcrest, is a rare resident in southern England. It also occurs as a passage migrant on the east coast, and some of these birds move inland and overwinter here (especially in the south-west).

Length 9cm

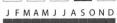

| 1 | 2 | 3 | **4** | 5 |

J F M A M J J A S O N D

How to find

■ **Timing** Although they are present in some areas all year round, Firecrests are more numerous and easier to see in autumn and winter. When Goldcrests arrive in large numbers in autumn they usually bring a few Firecrests with them.

■ **Habitat** Firecrests may be found in various kinds of wooded habitat. They show less of a strong affinity to conifers than Goldcrests do and will breed in wholly deciduous woodland. In winter they will use areas that are dominated by scrub rather than trees, often on or near the coast or in sheltered river valleys that stay a degree or so warmer than surrounding habitat. They are often found in company with Goldcrests and sometimes tits.

■ **Search tips** As with Goldcrests, finding Firecrests often comes down to listening for the song and call. Its vocalisations are very similar to those of Goldcrest but the song is simpler and the calls are lower-pitched. Whenever you encounter a flock of Goldcrests in autumn or winter, search carefully and check each bird, especially if you hear a call that sounds different to the rest.

Super sites

1. North Norfolk coast
2. Thetford Forest
3. Isles of Scilly
4. Porthgwarra
5. Portland Bill
6. New Forest
7. RSPB Dungeness

WATCHING TIPS

Much of the advice about watching Goldcrests also applies to Firecrests. It is also important to note any sign of breeding activity from Firecrests away from the regular breeding range, as this species has expanded its range and increased in numbers quite significantly since the turn of this century. As with Goldcrest, it is vulnerable to bad winter weather and a succession of harsh winters could have dire consequences for the still small and fragile breeding population. In the breeding season you could see the male's courtship display, in which the bright crown feathers are raised and spread to stunning effect.

Spotted Flycatcher

Muscicapa striata

A rather plain but elegant bird, the Spotted Flycatcher is a summer migrant, found throughout the UK and Ireland. Its population density varies considerably throughout its range, and it has undergone a steep decline since the late 20th century.

Length 14.5cm

1	2	3	4	5

J F M A M J J A S O N D

How to find

■ **Timing** Spotted Flycatchers are among the latest of our common summer visitors, with few seen before May, and many not on their territories until mid-May. They will have departed by the end of September.

■ **Habitat** These flycatchers nest in tree crevices (or suitable nest boxes), and feed on flying insects. Habitats that meet their requirements include deciduous woodland with sunny rides or glades, parkland with scattered mature trees, cemeteries and larger gardens. Passage migrants may briefly use treeless but scrubby environments on the coast.

■ **Search tips** This bird is usually found by sight rather than sound. Its feeding habits draw attention – it sits on an exposed low perch (often a bare twig in the lower parts of a tree) and from there makes fly-catching sorties, returning to the same perch to eat its catch. Its agile looping flights after insects are eye-catching. Look out for this behaviour wherever there are sunlit patches around mature trees, and a good population of flying insects. The call and song are both rather quiet and simple, easily lost among the voices of other woodland birds.

WATCHING TIPS

It is a real birdwatching treat to watch a Spotted Flycatcher plying its trade in a sunlit glade. As long as you keep a fair distance back you should be able to watch it make repeated fly-catching sorties, and even hear the snap of its bill as it takes its prey. Falling numbers of large flying insects may be linked to the decline of this species. If you have a good-sized country garden with trees, put up some open-fronted nest boxes to encourage this species, and ensure you use natural gardening methods to support a good insect population.

Ficedula hypoleuca   # Pied Flycatcher

A beautiful and charming summer visitor, the Pied Flycatcher is quite widespread in northern and western parts of Great Britain, though is very scarce in Ireland. It also occurs as a passage migrant on the eastern side of England.

Length 13cm

1	2	3	4	5

J F M A M J J A S O N D

How to find

■ **Timing** Pied Flycatchers occupy their breeding grounds in late April or early May, and may be seen on migration just prior to this. They disperse after breeding in late summer, and the autumn passage (much more pronounced than in spring) is through September. When there are 'falls' of migrating songbirds on our east coasts in autumn after easterly winds, this species is usually among those present.
■ **Habitat** This flycatcher is a classic bird of mature upland oak woodland with steep hills and valleys, although it will also use lowland and mixed forest in some areas. On migration it will stop off anywhere where there are trees or scrub.
■ **Search tips** Despite the male's bold patterning, this bird can be hard to spot among the light and shadows of woodland in full leaf. The males in fact may stand out more on dull rather than sunny days. The sweet simple song gives away a male on territory. When walking in the woods, look high in the canopy from time to time, and if you notice a hole-fronted nest box, wait a few minutes (a good distance back) to see if a flycatcher family is in residence.

Super sites

1. RSPB Wood of Cree
2. RSPB Haweswater
3. RSPB Lake Vyrnwy
4. RSPB Gwenffrwd-Dinas
5. RSPB Coombes and Churnet
6. RSPB Nagshead

WATCHING TIPS

This bird is generally more difficult to watch than the Spotted Flycatcher because it tends to fly-catch higher in the canopy, and spends more time picking insects from the foliage. Many nature reserves where the species occurs put up numerous nest boxes, and positioning yourself where you can watch an occupied box without disturbing the occupants is a good way to spy on Pied Flycatcher family life. After breeding the males moult into a dull, female-like plumage – something to be aware of if you are looking for passage migrants in autumn.

Bearded Tit

Panurus biarmicus

This charming small bird, tit-like in appearance but actually not closely related to the 'true tits', is an uncommon species in Britain, found all year in marshy areas mainly in southern England. Most populations are at or near the coast.

Length 14–15.5cm

| 1 | 2 | 3 | 4 | 5 |

J F M A M J J A S O N D

How to find

■ **Timing** Bearded Tits are on their breeding grounds all year, though are perhaps easiest to see in late summer and autumn when adults and young birds are moving around as family groups. They stay low in the reeds on windy days. In cold winters they may abandon more inland areas and head for the coasts where temperatures stay slightly higher.

■ **Habitat** This bird feeds and breeds in large reedbeds, and is unlikely to be seen anywhere else, though in winter it may turn up on smaller reedbeds that will not subsequently be used for breeding.

■ **Search tips** Pick a fine, still day and head for a coastal nature reserve with extensive reeds. The birds are easier to spot from an elevated path or hide with a view across the reeds, and choose a route where the sun is behind you as you look across the reeds. Look out for a small, long-tailed bird flying low over the reed tops or on muddy ground at the bases, and listen for the very distinctive dry tapping 'ping' call – feeding birds call constantly, and with patience and small adjustments of your viewpoint you should spot them.

Super sites

1. RSPB Leighton Moss
2. RSPB Newport Wetlands
3. RSPB Titchwell
4. RSPB Strumpshaw Fen
5. RSPB Minsmere
6. RSPB Elmley Marshes
7. RSPB Dungeness

WATCHING TIPS

Bearded Tits are delightful birds and will sometimes give wonderful views, although on good sites they do have a large area of reeds in which to lose themselves. Watching from a hide can give spectacular close views – or nothing at all. At some nature reserves, grit trays are provided for the birds (they need to eat grit to help break down the seeds in their diet) and in autumn and winter a succession of birds should visit them, giving good views (but don't stand too close). Look out for Bearded Tits in even quite small reedbeds in winter, as youngsters disperse to find new territories.

Aegithalos caudatus

Long-tailed Tit

This delightful small bird, found throughout the UK and Ireland except the far north of Scotland, is one of our most distinctive garden species. It is not closely related to the 'true tits' but is very like them in general behaviour.

Length 14cm

| 1 | 2 | 3 | 4 | 5 |

J F M A M J J A S O N D

How to find

■ **Timing** Long-tailed Tits are resident and easy to see at all times of year except mid-spring, when nesting pairs become quite discreet for a few weeks.
■ **Habitat** Any habitat with trees and bushes will support this species. Woodland, scrub, parkland, hedgerows and gardens are all good places to look.
■ **Search tips** Feeding flocks of Long-tailed Tits keep up a constant exchange of ticking and purring contact calls, which is usually the first clue of their presence. Watch the trees or bushes from which they are calling and you should soon spot the birds, picking their way through the twigs or flying from one tree to the next. They will feed at all heights, although rarely on the ground. In the breeding season they can seem to completely disappear, but pairs do continue to call. The nest is furnished with a huge number of small soft feathers, so if you find a quantity of feathers (perhaps the remains of a Sparrowhawk's or other predator's kill) in spring, you may see Long-tailed Tits coming to collect them. In winter, small numbers of the white-headed Scandinavian subspecies may visit Britain.

WATCHING TIPS

Most gardens will attract Long-tailed Tits from time to time and they will use bird feeders, although perhaps less readily than other tits. Fat blocks in cage feeders seem particularly attractive. These gregarious birds like to eat communally, and often seem to appear out of nowhere and cover a feeder for a few minutes, before moving on and perhaps not returning for the rest of the day. When you encounter a feeding party of Long-tailed Tits while out walking, wait ahead of the way they are moving and they may completely surround you, showing no fear whatsoever.

Blue Tit

Cyanistes caeruleus

This colourful little bird is a very common and popular garden visitor, familiar to everyone who puts out bird feeders in even the smallest gardens. It is found throughout the UK and Ireland except the northernmost Scottish islands.

Length 11.5cm

1	2	3	4	5

J F M A M J J A S O N D

How to find

■ **Timing** Blue Tits can be seen all year round with ease. They make most use of garden feeders in the winter months, and newly formed pairs will be inspecting nest boxes in late winter to begin nest-building in March.

■ **Habitat** Anywhere with trees will have Blue Tits – they are only scarce in very open, exposed environments. Suitable woodland may be upland or lowland, damp or dry, dense or patchy. They prefer deciduous woodland but will use coniferous forest, particularly native pine woods in the Scottish Highlands. They are common in parkland, copses on farmland and in gardens.

■ **Search tips** Blue Tits are very active, quite approachable, and call frequently – all factors which make them very easy to find. When a small agile bird moving in the treetops attracts your attention, the chances are good that it will turn out to be a Blue Tit. In winter it may roam in flocks, often with other tit species in tow. It is probably the likeliest species to be using a hole-fronted nest box, either in a garden or on a nature reserve.

WATCHING TIPS

You don't even need a garden to enjoy watching Blue Tits from your home, as they will readily come to feeders attached to the window glass. Sunflower seed hearts and fat balls are safe food options to offer all year – in mid-summer you may attract numerous newly fledged chicks, which quickly learn from their parents how to take food from hanging feeders. This species readily uses nest boxes and will also roost in them through the winter months – choose a box with a fitted camera for a real insight into the behaviour of this species.

Parus major

Great Tit

The largest of our tit species, this is a boldly colourful bird and a very familiar visitor to garden feeding stations. Like the Blue Tit, it occurs throughout the UK and Ireland, but not in the northernmost Scottish islands.

Length 14cm

1	2	3	4	5

J F M A M J J A S O N D

How to find

■ **Timing** Great Tits are easy to see throughout the year. Males begin singing in February, and nesting is well under way in March. Fledged chicks appear in June. As natural food dwindles through late autumn, numbers visiting garden feeding stations begin to climb.

■ **Habitat** The Great Tit is a woodland bird by adaptation, and occurs in all kinds of habitats with some trees. It needs tree holes in which to nest so does best in areas with mature trees but will readily use nest boxes and other artificial hollows, including improbable sites like postboxes, broken drainpipes and even cigarette bins.

■ **Search tips** Great Tits will visit nearly any garden, even in city centres, if food is put out, and will also come to window feeders. In the wider countryside they are quite noticeable, feeding at all heights, climbing tree trunks to investigate cracks and hollows, and coming down to take fallen beechmast on the ground. They call often, and in early spring the male's loud, ringing two-note song is one of the most obvious woodland bird songs.

WATCHING TIPS

At the feeding station, Great Tits use their greater size to bully and dominate the smaller tit species. They also have the potential to win disputes over ownership of nest boxes, but are excluded from those with smaller holes – put up boxes with a 28mm hole to attract this species. Don't neglect the feeding station in summer – the fledglings will appreciate having an easy food source as they learn to forage for themselves – but avoid putting out whole peanuts through spring and summer. Fat-based foods are ideal to offer all year round, as are sunflower seeds.

Crested Tit

Lophophanes cristatus

Although the Crested Tit is common and widespread in all kinds of wooded habitats on the near continent, in Britain it is only found in the coniferous forests of the Scottish Highlands. It is fairly common in this small range.

Length 11.5cm

| 1 | 2 | 3 | 4 | 5 |

J F M A M J J A S O N D

How to find

■ **Timing** Crested Tits stay on their breeding grounds all year round. They may be easier to see in winter when they form small flocks and roam more widely – this is also the time when they are most likely to visit gardens.

■ **Habitat** The best habitat for this species is the native Caledonian pine forest that still remains in parts of the Highlands, such as at the RSPB's Abernethy Forest reserve in Strathspey. It will also use commercial pine plantations, and will readily visit gardens adjoining areas of pine wood.

■ **Search tips** When walking in suitable woodland, keep checking the trees for small birds moving in typical agile tit fashion among the pine twigs. Note that Blue, Great and especially Coal Tits are likely to also be present, and often (especially outside the breeding season) there will be mixed flocks around. The pointed crest is noticeable from a distance and when the bird is in silhouette. The Crested Tit's call, a soft and rather high-pitched down-slurring trill, is distinctive and stands out among the calls of the more common tit species.

Super sites

1. RSPB Culbin Sands
2. RSPB Corrimony
3. Rothiemurchus
4. Aviemore
5. RSPB Abernethy Forest

WATCHING TIPS

Crested Tits are engaging to watch and have little fear of humans, although the shady and permanently leafy environment of the pine wood can make them difficult to see clearly. Visiting the woods after a winter snowfall can actually brighten up the scene, and also guarantees plenty of activity from the birds as they will need to feed throughout the short daylight hours. One or two Crested Tits are often found among mixed feeding flocks made up primarily of Coal Tits. If you live (or holiday) in Crested Tit country put out feeders and you may be rewarded with great views – they will also use nest boxes.

Coal Tit

This small tit species falls into third place behind Blue and Great Tits in terms of its frequency as a garden visitor. It is widespread and common throughout nearly all of the UK and Ireland in wooded areas (especially coniferous).

Length 11.5cm

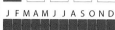

| 1 | 2 | 3 | 4 | 5 |

J F M A M J J A S O N D

How to find

■ **Timing** You can see Coal Tits all year round in suitable habitat. They are most likely to visit garden feeders when natural food is in short supply, so between late autumn and early spring, and fledglings will also use feeders in mid-summer.

■ **Habitat** Any kind of woodland can support Coal Tits but their numbers are especially high in coniferous forest of all kinds, including quite closed spruce plantations that support few other birds. They also visit gardens, although are more likely to come to larger gardens in more rural surroundings.

■ **Search tips** Coal Tits are typical tits in their behaviour, climbing among twigs, dangling upside down and constantly active. This is noticeable in itself, but also listen for the calls, which are higher pitched than those of the commoner tits and lack the harsher trilling or churring notes, and the high, rapid, two-note song. In pine woodland this is usually the most common tit species around. Coal Tits in the garden tend to make flying visits, taking a seed or nut away rather than eating it on the feeder.

WATCHING TIPS

This tit, smaller and shyer than Blue and Great Tits, is dominated and bullied by both at garden feeders, so you may need to watch for a while to spot one. It has the habit of caching food – you may notice it flying away with a nut and carefully hiding it in a bark crack or similar hiding place. Coal Tits will use nest boxes – if you have a local population, the more boxes you put up the better your chances as the local Blue and Great Tits will outcompete the Coal Tits. This species is sometimes confused with Marsh Tit, but with good views its large bib and double wingbar should eliminate doubt.

GARDEN BIRDS

If you have a garden, you're perfectly placed to hone your bird-finding skills on your own turf, and you can also provide a miniature wildlife haven that makes a real difference to your local biodiversity. Even small urban gardens have the potential to attract 10 or more different species, while avid 'garden-watchers' in particularly favourable parts of the country have a garden bird list that reaches well into three figures.

How to watch

For many of us, garden birdwatching is a casual pursuit, something to do while washing up or mowing the lawn. To see more birds you'll need to put in more time, but the kitchen window is a good starting point. Keep binoculars by the window that you look out of most often, and position your feeding stations where you can see them from this window.

Taking things a step further, if you have a large garden, you could set up a second feeding station and drinking area and set up a hide from which you can view it. You may have a perfectly placed garden shed, or you could invest in a simple tent-like hide. This will enable you to get much closer to your garden birds, ideal if you want to take photographs or make detailed observations of behaviour.

Eyes to the skies

People who record an impressive range of species on their garden lists don't limit themselves to birds that actually 'touch down' in their gardens but also count 'flyovers'. Putting aside some time to watch the sky (perhaps from a comfortable reclining chair!) will quickly boost your list and also help hone your identification skills of birds in flight. Learning the outlines, calls and flight styles of common birds will put you in good stead when something less common comes along.

If you live near the coast, on a hilltop or along a natural flyway like a river valley, you should record a great variety of species. Large numbers of birds are on the move in spring and autumn, especially early and late in the day, while midday is a good time to look out for birds of prey.

A home for birds

Making your garden hospitable to birds means supplying the resources they need. If you want Blue Tits to visit your garden in winter, offering a feeder of peanuts or sunflower hearts will do the trick. If you want them to nest, you'll also need to supply a nest box. However, there's more to encouraging garden birds than putting out feeders and boxes – you also need to take care of the biodiversity of your garden. The parent Blue Tits need to find huge numbers of caterpillars to sustain their growing chicks, but if your garden has no native plant species or is covered with pesticides it will not hold enough of this natural food.

When planning and managing your garden with birds and other wildlife in mind, think native plants, organic pest control, and structural variety. Our native insects are adapted to live on our native plants – non-native plants support far fewer insects. Aggressive chemical pest control kills off birds' food supplies, so deal with pests conservatively and let nature take care of itself. Including a diversity of plant types and shapes provides the different kinds of cover birds need for nesting and feeding, and a wide choice of natural foods from buds and berries to nuts and seeds. Adding extras like a woodpile, drystone wall or pond creates more homes for invertebrates and so more feeding opportunities for birds.

Goldfinches, Siskins and Greenfinch

THE WINTER GARDEN

When the countryside experiences a spell of severe weather, gardens really come into their own as havens for both familiar and unfamiliar birds. If there is a freeze and some snowfall, keep feeders topped up, make sure you put out fresh water frequently, and keep your eyes open for unusual visitors.

In the winters of 2009–2010 and 2010–2011, many householders found countryside birds like Meadow Pipits and Skylarks, and waders like Woodcocks and Snipes, taking advantage of their garden comforts. You can help birds like this by keeping part of the ground swept clear of snow, and by putting out high fat, high protein foods. Mealworms are always good and appreciated by many species. Bird food manufacturers now produce suet-based foods in small pellets, sometimes with squashed insects incorporated, which can be scattered on the ground.

Blackbird

Willow Tit

Poecile montanus

This black-capped tit, very similar to the Marsh Tit, is becoming a rare bird in Britain. Its stronghold is in the north-west of England, and it is rather scarce elsewhere. It is absent from most of Scotland and from Ireland.

Length 11.5cm

| 1 | 2 | **3** | 4 | 5 |

J F M A M J J A S O N D

How to find

■ **Timing** Willow Tits remain on their breeding grounds throughout the year. Winter is the best time to try to see them at feeding stations – in the breeding season they can become very discreet until the chicks fledge.

■ **Habitat** In Britain, this is a bird of mainly damp and thick low-lying woodland, often with a high proportion of willow trees. It is much scarcer in dry and open forest. Where gardens adjoin suitable habitat it will readily visit garden feeding stations.

■ **Search tips** Your best chance of finding a Willow Tit is to visit a nature reserve with a good population of them and spend some time watching the feeding station. Otherwise, this tit may be found moving actively around the trees, picking small insects from the twigs and searching rotting tree or branch stumps. In winter, a small group may team up to maintain a feeding territory – they are not especially inclined to join mixed tit flocks. Listen for the call – a very distinctive, low-pitched nasal 'djerrr djerr'. The rather Wood Warbler-like song is not often heard but is also distinctive.

Super sites

1. RSPB Wood of Cree
2. RSPB Ken-Dee Marshes
3. RSPB Fairburn Ings
4. Denaby Ings
5. Potteric Carr
6. Cannock Chase
7. RSPB Sandwell Valley

WATCHING TIPS

Willow and Marsh Tits are perhaps the most problematic 'species pair' in Britain, especially as both become scarcer, giving birdwatchers fewer opportunities to study them. Much traditional advice on separating them has now been shown to be unreliable. The best feature is call, but also look for a more neckless 'egg shape' in the Willow tit, with more orangey plumage tones. There are also subtle differences in bill and cheek pattern. Documenting both species is very important, as the Willow Tit in particular is in serious need of active protection.

Like its close relative, the Willow Tit, this species is declining in Britain, but remains quite common and widespread in England and Wales. Its range just reaches southern Scotland, but it does not occur in Northern Ireland or Ireland.

Length 11.5cm

| 1 | 2 | 3 | 4 | 5 |

J F M A M J J A S O N D

How to find

■ **Timing** Marsh Tits can be found in the same areas throughout the year. Like other tits, they will visit garden feeding stations more often at times when natural foods are scarce, in between late autumn and early spring.

■ **Habitat** Despite its name, this species actually favours drier and more mature woodland. It uses deciduous woodland rather than coniferous, and as it is a hole-nester needs at least some mature trees within its habitat. Although it prefers extensive areas of tree cover it will also use patchier woodland and will readily visit rural gardens.

■ **Search tips** Marsh Tits are often first noticed when they give their sneezing 'pit-choo' call, which tells them apart from Willow Tits. They feed more often in the middle to lower levels in the wood – look out for them on the ground below beech trees in autumn when a good crop of beechmast has fallen. When visiting feeding stations they are often dash-and-grab merchants, taking seeds or nuts away to eat in peace without harassment from more dominant tit species. An uninterrupted hour watching your feeders should determine if one is visiting.

WATCHING TIPS

In most areas, a black-capped brown tit is more likely to be a Marsh than a Willow Tit. Marsh Tits have two distinctive features that are hard to spot on a moving bird but show up well in photos – a white spot on the inner upper mandible, and a distinctively two-toned cheek (white at the front, pale grey-buff at the back, while the cheek of the Willow Tit is all white). Even a poor photo can be enough to show these traits and help with identification. Marsh Tits will use nest boxes, and may prefer to take seed scattered on a bird table rather than use a hanging feeder.

Nuthatch

Sitta europaea

This is a very distinctive small woodland bird. It is common and widespread in wooded parts of England and Wales, and is extending its range north with new populations in southern Scotland. It is absent from Ireland and Northern Ireland.

Length 14cm

| 1 | 2 | 3 | 4 | 5 |

J F M A M J J A S O N D

How to find

■ **Timing** Nuthatches can be seen all year in the same areas, but become quiet and discreet when nesting. As with most other nut- and seed-eating garden birds, their presence in gardens increases between late autumn and early spring.

■ **Habitat** As long as there are reasonable numbers of mature deciduous trees around, most areas in town and country will have Nuthatches. They are just as likely to be encountered in open parkland as in dense woodland. They need tree holes for nesting (and are able to adapt holes that are too large to their needs) and are common visitors to bird tables and feeders.

■ **Search tips** This bird is often mistaken for a woodpecker at first glance as it clings to tree trunks and branches in a similar fashion. However, it moves with more freedom than a woodpecker. If you don't notice the moving bird you may well hear its loud ringing 'tuit tuit' call, or hear regular tapping sounds as it hammers away at a nut wedged in a bark crack. When it comes to a hanging feeder, it will often scare off other small birds.

WATCHING TIPS

Nuthatches are entertaining birds to watch. They move with great agility in the trees, climbing down trunks head first just as easily as climbing up, and show similar skill on the bird feeders. Heavy-billed, they are able to break open quite hard nuts and seeds, but do not excavate their own nest-holes. However, they will narrow the size of a hole's entrance by plastering its edges with mud – nest boxes used by Nuthatches have been found to contain an amazing amount of 'plaster' on the inner walls. When nesting they seem to vanish from the woodlands, but in early summer you may spot fledglings quietly waiting in the trees for their parents to bring food.

Certhia familiaris — Treecreeper

Certhia familiaris

Treecreeper

This bird is a common inhabitant of woodlands throughout the UK and Ireland, only missing from the most northerly Scottish islands. Despite this it is not the easiest bird to see and watch as it has rather unobtrusive ways.

Length 12.5cm

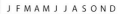

1	2	3	4	5

J F M A M J J A S O N D

How to find

■ **Timing** You can see Treecreepers in woodlands at any time of year – their behaviour does not change much through the seasons although like most birds they are hard to find when nesting. In winter they may use sheltered nooks and crannies (including nest boxes) as roost sites.

■ **Habitat** This species requires habitats with a reasonable amount of trees, as it feeds almost exclusively on small insects picked out from cracks in the bark. The woodland may be dry or damp, deciduous, coniferous or mixed, but it is not normally found in young plantations.

■ **Search tips** You will often spot a Treecreeper when you have paused to look at another bird. It is almost invariably seen on a tree trunk, working its way up from the base in a spiralling manner (though if it sees you watching it may quickly move to the far side of the trunk). You may also spot it as it flies from near the top of one tree to the base of another. It has a high-pitched, Goldcrest-like call, and in spring gives a similarly thin and high descending and twittering song.

WATCHING TIPS

There can be few British birds with a more simple and straightforward way of life than the Treecreeper. When it has climbed to near the top of one tree it will usually start again at the very bottom of another quite nearby and this is when it is best seen. It may sometimes scale a brick wall or wooden fence instead of a tree. It will visit garden trees and occasionally feeders (choosing fat blocks rather than seed), and if you have a redwood tree in your garden you may find a Treecreeper roosting in a hollow in its soft bark.

Golden Oriole

Oriolus oriolus

This stunning bird is a very rare breeding species in Britain, and an uncommon passage migrant. The only publicised regular breeding site is in west Norfolk, while passage migrants are most often encountered on the south and east coasts.

Length 24cm

| 1 | 2 | 3 | **4** | 5 |

J F M A M J J A S O N D

How to find

■ **Timing** Look for passage migrants in late spring and again in early autumn, in any coastal area with mature trees. Breeding birds are on territory from mid-May and will be singing up until mid-July.

■ **Habitat** Migrants will make use of any wooded or scrubby habitat. The breeding birds in Norfolk stick to an area of riverside black poplar trees.

■ **Search tips** Although males are startlingly colourful birds, both sexes are very difficult to see and you will probably need plenty of patience to get any kind of sighting. If you visit RSPB Lakenheath Fen between mid-May and mid-July you do stand a good chance of at least hearing the male's song, which is a remarkable series of loud, fluty, whistling notes that sounds as though it belongs in a tropical rainforest. Visit in the early morning for your best chance of a sighting, and scan repeatedly among the poplars from the viewpoint. Searching for migrants on the coast is also best done in the morning

Super sites

1. RSPB Lakenheath Fen
2. Isles of Scilly

WATCHING TIPS

If you put in a fair amount of time you should eventually enjoy reasonable views of Golden Orioles at RSPB Lakenheath Fen. Early in the breeding season may be the best time for interesting sightings, as the birds will be pairing up and establishing territories, and you could see territorial chases through the trees. If you find a singing bird away from this site, notify your local recorder, as breeding does occasionally occur elsewhere. While the males are unmistakable, a poorly seen female could be confused with a Green Woodpecker, likely to be present in the same habitat..

Lanius collurio

Red-backed Shrike

Once quite a widespread breeding bird in Britain, the Red-backed Shrike is now only a sporadic breeder in tiny numbers, though could yet recolonise given enough protection. It also occurs as a regular passage migrant in small numbers.

Length 17cm

| 1 | 2 | 3 | **4** | 5 |

J F M A M J J A S O N D

How to find

■ **Timing** Red-backed Shrikes are seen on migration in May and again in September and October. Sightings between these months could indicate breeding activity. Arrivals on the coast in autumn often occur during or straight after periods of easterly winds.

■ **Habitat** This bird likes open, scrubby, sunny habitats with an abundance of large insects (which make up the bulk of its diet). It may be found on heathland, downland, quiet mixed farmland with hedgerows and neglected orchards. On passage it is usually found in coastal scrub.

■ **Search tips** Like all shrikes, this bird hunts mainly from an exposed lookout perch, so can be quite obvious even from a distance as it waits on top of a bush or post in a quite upright stance. Migrants rarely call (you may hear a series of hard 'tak' calls from alarmed birds) but males holding territory have a fast twittering or warbling song, which they give early in the breeding season. Discovery of a 'larder' of insects and other small creatures impaled on thorns is a sure sign of a shrike's presence.

WATCHING TIPS

This bird is like a miniature raptor, and is interesting to watch as it hunts and deals with its prey. Insects may be caught in a pursuit flight or pounced on from a perch – it will also hover briefly before dropping onto prey. Breeding has occurred in England, Scotland and Wales so far this century so could happen anywhere – be alert whenever walking in scrubby countryside in summer, and inform your recorder if you see evidence of nesting. The species has long been a favourite target of egg thieves, so needs strict protection. Migrating males may sing and hold territory briefly but will move on if no female arrives.

Great Grey Shrike

Lanius excubitor

This handsome and unmistakable bird of open countryside is a rare winter visitor to Britain. It is most often encountered in the south-east but could occur anywhere, and certain sites regularly hold a single bird year after year.

Length 24–25cm

1	2	3	4	5

J F M A M J J A S O N D

How to find

■ **Timing** Great Grey Shrikes begin to arrive in late September and may stay to the end of March or even into April. Mid-winter is probably the likeliest time to find one. Numbers overall seem to be higher in colder winters.

■ **Habitat** Open countryside with scrubland attracts this species, which likes a wide-ranging view when on the lookout for prey. Farmland with hedgerows, downland with scattered bushes and lowland heath are all good places to look.

■ **Search tips** When in good habitat, scan the tops of all bushes and small trees, and along lines of fence posts and the fencing itself, looking out for an upright, long-tailed bird which often looks very white from a distance. This is a very solitary species in winter, and its feeding territory can be quite sizeable, so if you're searching for a reported bird, make sure you include quite a wide area in your search, and look well ahead of your position – a telescope will be very helpful. Look out for mobbing or avoidance behaviour from small birds as well, as they may harass a perched shrike or flee from one in flight.

Super sites

1. Westleton Heath NNR
2. RSPB Aylesbeare Common
3. Woodbury Common
4. RSPB Arne
5. New Forest
6. Thursley Common
7. Ashdown Forest

WATCHING TIPS

The Great Grey Shrike is usually quite easy to watch once located, as it perches and hunts out in the open. It is unlikely to tolerate close approach, though. It is a true predator, taking a fair proportion of small mammals and even other birds as well as what insects it can find – watch for a while and you could see its hunting behaviour. Like the Red-backed Shrike, it stores prey in larders, using thorns or barbed wire spikes to impale food items for later consumption. When hunting it chases prey in the air and on foot and may hover expertly like a Kestrel before pouncing.

Garrulus glandarius

Jay

A colourful member of the crow family, the Jay is common throughout most of the UK and Ireland, becoming scarcer further north. It is absent from the north of Scotland, and rather localised in Northern Ireland and Ireland.

Length 34–35cm

| 1 | 2 | 3 | 4 | 5 |

J F M A M J J A S O N D

How to find

■ **Timing** Jays are on their breeding grounds all year round, although in some years there is an influx of extra birds from the continent in autumn and winter. It is probably easiest to see in autumn, when it is very actively collecting and storing acorns for the winter.

■ **Habitat** Primarily a bird of mature deciduous forest with a reasonable under-storey, the Jay may be encountered in any partly wooded habitat including orchards, farmland copses and town parks. As long as there are trees around, Jays will also readily visit larger gardens.

■ **Search tips** The Jay is the woodland bird that will see you before you see it and immediately warn all other birds in the vicinity with a loud, harsh screaming call. When you hear the call, look out in the high branches of trees for a large but discreetly moving bird, and look out for the flashing white rump of a departing Jay. In autumn, when walking or driving in the countryside look out for Jays flying steadily overhead carrying acorns – the flight silhouette shows strikingly round-ended wings and tail.

WATCHING TIPS

Jays are normally very shy and alert, and quick to disappear if you are too close, so trying to get a good view of them can be frustrating. In town parks they may become more confiding, although very rarely are they anything like as approachable as the other common corvids. Setting up a woodland hide, perhaps overlooking a pond, may help give you good views. If Jays live in the vicinity of your garden, you may be able to attract them by providing food like soaked dog biscuits on the lawn edges, close to the safety of trees or bushes.

Magpie

Pica pica

Familiar and unmistakable, the Magpie is a very common British bird, absent only from north Scotland.

Length 44–46cm

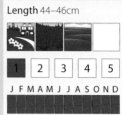

| 1 | 2 | 3 | 4 | 5 |

J F M A M J J A S O N D

How to find it

- **Timing** Magpies are easy to see at all times of year.
- **Habitat** Most habitat types can attract Magpies, as they are true opportunists which will roam considerable distances while on the lookout for food. They need tall mature trees in which to nest, so wooded areas hold the highest numbers, but they are also common in towns and on roadsides around road kill.
- **Search tips** With its striking plumage, bold ways and repertoire of loud calls, the Magpie is very easy to find. Food placed out on the lawn or on a bird table will probably attract it to your garden.

WATCHING TIPS

In good light, the Magpie is a dazzlingly beautiful bird with violet and emerald iridescence to its black wing and tail feathers. It is intelligent and versatile, and any time spent watching a feeding Magpie will probably reveal aspects of its enterprising and inventive habits.

Jackdaw

Corvus monedula

The smallest of the black crows, the Jackdaw is a common bird in the UK and Ireland, except for the far north-west of Scotland.

Length 33–34cm

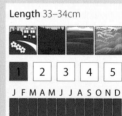

| 1 | 2 | 3 | 4 | 5 |

J F M A M J J A S O N D

How to find it

- **Timing** Jackdaws can be seen all year round with ease. They form large flocks through the winter months.
- **Habitat** These birds are present in town and countryside alike, as they will nest in both mature trees and in crevices of buildings. They forage on open fields, both pasture and arable, often with other corvids.
- **Search tips** Most towns and villages have breeding Jackdaws – look for them perched on chimneys or swirling in flocks overhead. They have a more compact shape and agile flight than Carrion Crows. They will often visit gardens to take food left in open places.

WATCHING TIPS

Like all crows, Jackdaws are intelligent, inquisitive and interesting to watch. They are gregarious all year (where habitat permits they nest colonially) so social interactions may be observed throughout the year. They are skilled fliers, and may perform mid-air chases and other aerobatics.

Chough

This is the scarcest species of crow to occur in Britain. It is found on the coasts of Wales and south-west Scotland, the Isle of Man and Cornwall, but is most numerous along the west coasts of Ireland and Northern Ireland.

Length 39–40cm

1	2	3	4	5

J F M A M J J A S O N D

How to find

■ **Timing** Choughs can be seen year-round at their preferred breeding sites, with juveniles possibly wandering further afield in autumn, and flocks forming in winter.

■ **Habitat** The Chough is more or less restricted to coastlines with quite high cliffs, nesting on ledges or in crevices within the rock. It needs well-grazed fields of short turf with some bare ground on the clifftops or just inland.

■ **Search tips** When in the right habitat, scan clifftop fields for parties of feeding Choughs, and look out for them flying overhead. The outline in flight is distinctive and unlike that of any other crow, even if the long curved bill cannot be made out – the wings have very broad and pronounced 'hands' with long, well-separated flight feather 'fingers'. The drawn-out onomatopoeic call is also noticeable. After a long period of decline in Britain, Choughs are now on the increase, and there is plenty of potential for them to colonise new areas, so be alert to the possibility of Chough sightings on the coasts away from regular areas.

Super sites

1. RSPB Loch Gruinart
2. RSPB The Oa
3. RSPB Rathlin Island
4. The Dingle Peninsula
5. Calf of Man
6. RSPB South Stack
7. Strumble Head
8. RSPB Ramsey Island
9. The Lizard

WATCHING TIPS

These lovely birds are masters of the air and thrilling to watch in flight as they lift and tumble on the wind. Most good sites for Choughs have the added advantage of spectacular scenery, making these birds well worth a special trip if you don't live close to a breeding site – and many Chough breeding grounds also hold colonies of other cliff-nesting birds, including Ravens and Peregrine Falcons all year round. Any sightings of Choughs away from the breeding grounds should be passed on to local recorders, but note that the species is very unlikely to be found inland.

Rook

Corvus frugilegus

A large, all-black crow, the Rook is a very common species throughout the UK and Ireland, only absent from the far north-west of Scotland. It is often confused with the Carrion Crow but with good views there are several differences.

Length 45–47cm

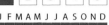

| 1 | 2 | 3 | 4 | 5 |

J F M A M J J A S O N D

How to find

■ **Timing** Rooks are easy to find all year. They occupy their breeding colonies from late winter until mid-summer.

■ **Habitat** Colonies are usually in quite tall trees, sometimes in isolated stands but more often on the edges of more extensive woodlands. The birds forage in nearby farmland fields, and will also come to road kill, scavenge scraps from motorway service stations, and visit larger gardens, but are much less likely to be found in truly urban areas than are Carrion Crows.

■ **Search tips** A rookery is most obvious in winter, when the collection of large untidy nests is obvious among the bare branches, and in late winter there will be much noisy activity around the nests from the birds themselves. At all times of year, you should be able to easily find Rooks flying overhead and settled on pasture fields where they search for insects and other invertebrates – by mid-summer the adults will be joined by fledged young. They are highly gregarious at all times, so when you see one there will probably be many more nearby.

WATCHING TIPS

Nest-building, mating and other breeding activity among Rooks is quite easy to observe, as they are early breeders and the season is well under way before the leaves have grown in. A telescope will be helpful as the nests are usually quite high up. Young Rooks have fully feathered faces and so look much more like Carrion Crows than their parents do, but with careful observation you will notice the differences – as well as the fact that the young Rooks will be with Rook flocks. These corvids are highly intelligent, and much inventive feeding behaviour has been documented among wild birds.

Corvus corax

Raven

Approaching the size of a buzzard, this powerful bird is our largest crow and largest 'perching bird'. It is found mainly on high ground and coasts in the western half of Great Britain, and throughout Ireland.

Length 60–70cm

1	2	3	4	5

J F M A M J J A S O N D

How to find

■ **Timing** Ravens can be seen around their breeding grounds all year round. In autumn and winter dispersing young birds may turn up in other areas, and some upland pairs may move to lower ground for the winter.

■ **Habitat** This is a bird of rough, hilly ground with crags, and of rocky coastlines and cliffs. It does not often venture into urban areas, although may be seen over smaller towns that adjoin suitable habitat. It is attracted to any form of carrion, including road kill.

■ **Search tips** In suitable habitat you are most likely to spot a Raven on the wing. The very deep croaking call is far-carrying and will draw your attention to a very large but surprisingly agile and aerobatic black bird, which may well be flying with apparent enjoyment on days that are windy enough to keep most birds on the ground. When walking along sea cliffs, look out for Ravens wheeling around the cliff face. The bulky stick nest is usually built on a crag or cliff ledge, and is noticeable from a distance.

WATCHING TIPS

With a good view the Raven is an unforgettably dramatic-looking bird with its bulk and fearsome weapon of a bill. Reasonable views are all you need to identify it from the smaller and slighter Carrion Crow. In the air it is breathtaking to watch, soaring, gliding and tumbling on windy days. In many areas its habitat is shared with Peregrine Falcons, and the two species seem to have a mutually antagonistic relationship – mid-air skirmishes between these two equally skilled aerial masters are commonplace, and absolutely thrilling to watch.

Carrion/Hooded Crows

Corvus corone/C. cornix

Until recently, these two crows were regarded as different subspecies of the same species. Carrion Crows breed in England, Wales and the south-eastern half of Scotland, while Hooded Crows breed in the north-western half of Scotland and in Ireland.

Length 44–50cm
Wingspan 110cm

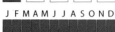

1	2	3	4	5

J F M A M J J A S O N D

How to find

■ **Timing** Carrion and Hooded Crows can be found in the same general areas all year round. Hooded Crows wandering from their usual range are most often found in winter.

■ **Habitat** These crows are habitat generalists, roaming and foraging in practically every habitat type including the seashore and city centres, but their key habitat is anywhere with tall trees in which they can nest, with open countryside nearby with access to bare or lightly turfed ground.

■ **Search tips** Both crow species are usually very easy to find in open farmland and in towns and villages, where they will often patrol village greens and other short turf. Hooded Crows are easy to recognise by their unique grey-and-black pattern, but Carrion Crows may be confused with other corvids. In flight they are noticeable for the very steady 'rowing' flight action, lacking the flair and agility of the Raven and Jackdaw. In the part of Scotland where distributions of the two forms meet, there is a 'hybrid zone' where intergrades between them frequently occur with variable amounts of grey plumage.

WATCHING TIPS

Carrion Crows are easiest to watch in town environments. In busy urban parks they will become almost as blasé about human presence as the ducks and squirrels, and their behaviour can be observed with ease. Hooded Crows are less likely to be found in urban settings, partly because their range has a lower proportion of town compared to country, but can easily be watched feeding in fields and on the seashore. It is often stated that these crows are solitary, a feature that distinguishes them from Rooks, but large feeding or roosting flocks are found in some areas.

Sturnus vulgaris

Starling

Although it has declined sharply since the late 20th century, this remains one of our most common, widespread and familiar birds.

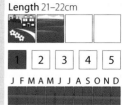

Length 21–22cm

| 1 | 2 | 3 | 4 | 5 |

J F M A M J J A S O N D

WATCHING TIPS

With their jaunty, quarrelsome characters and gregarious habits, Starlings are popular garden visitors. Offer suet balls and fruit to encourage them. They are also likely to nest in any hole in your roof or wall, and will use a large hole-fronted nestbox.

How to find it

■ **Timing** Starlings are found in most areas all year round. Spectacular large roosting flocks can be seen between late autumn and late winter.
■ **Habitat** It is common in both town and country, and will breed on woodland edges. Large winter roosts form within reedbeds or around tall buildings.
■ **Search tips** On autumn and winter afternoons, look out for flocks of Starlings flying restlessly in wide circles – this could help you find a major roost site, where you can enjoy an 'air show' as thousands of birds prepare to settle for the night.
■ **Super sites 1.** RSPB Leighton Moss, **2.** Aberystwyth Pier, **3.** RSPB Ham Wall, **4.** Brighton seafront.

Pastor roseus

Rose-coloured Starling

This is a nomadic species of central Europe and Asia, and an unpredictable rare visitor to mainly western parts of Britain.

Length 21–22cm

| 1 | 2 | 3 | 4 | 5 |

J F M A M J J A S O N D

WATCHING TIPS

A few very lucky householders have enjoyed the sight of a stunning adult Rose-coloured Starling visiting their garden. The dull juveniles must often go overlooked, but you are more likely to see one than an adult. In a garden setting they will often be quite confiding and approachable.

How to find it

■ **Timing** The peak months for sightings are September and October, with mainly juvenile birds involved.
■ **Habitat** When Rose-coloured Starlings occur in Britain, they are usually seen in Starling flocks. Many sightings are in gardens and suburban environments in general, otherwise hedgerows and farmland fields are good places to try.
■ **Search tips** It is always worth checking through Starling flocks for this species. Adult birds are unmistakable. Most Rose-coloured Starlings are still in pale brown juvenile plumage when they are found, while juvenile Starlings in September and October have already partly or completely moulted into their white-spotted black first-winter plumage.

Tree Sparrow

Passer montanus

In the last century this sparrow suffered one of the most severe declines of any British species. It is present but localised across much of central and eastern Great Britain, but is only found on the east side of Ireland.

Length 13–14cm

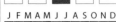

1	2	3	4	5

J F M A M J J A S O N D

How to find

■ **Timing** Tree Sparrows are on their breeding grounds all year, but in winter feeding flocks may wander more widely. They are most likely to be found in gardens between October and March.

■ **Habitat** An insect-eater in summer and seed-eater in winter, the Tree Sparrow does best in habitats with an abundance of weedy plant growth. It also needs tree holes or similar cavities for nesting. Deciduous woodland edges adjoining mixed farmland with mature hedgerows, areas of set-aside and stubble fields is ideal. It visits garden feeding stations, especially in winter, and will use hole-fronted nest boxes.

■ **Search tips** Tree Sparrows can be hard to find in the general countryside, but on nature reserves where their specific needs are met they can do very well and form sizeable breeding colonies. Where nest boxes are provided, they will be in use in spring and summer, and you may see the males calling from on top of the boxes. In winter look for them around countryside feeding stations and among other seed-eating songbirds flocking in weedy and stubble fields.

Super sites

1. RSPB Loch of Strathbeg
2. RSPB Portmore Lough
3. North Slob
4. RSPB Fairburn Ings
5. RSPB Bempton Cliffs
6. RSPB Blacktoft Sands
7. RSPB Old Moor
8. Rutland Water
9. RSPB Freiston Shore
10. RSPB Dungeness

WATCHING TIPS

These attractive birds make a very special addition to the garden bird table. Gregarious and boisterous, they are similar to the more familiar House Sparrow in general behaviour, and like that species enjoy bathing in dusty patches as well as water – if they visit your garden, offer both facilities if possible. It is also worth providing several nest boxes (with a 28mm hole), positioned a couple of metres above ground, if you have a local population in spring and summer. Note that, unlike with House Sparrow, the sexes look the same.

House Sparrow

Passer domesticus

This is one of our commonest and best known birds, found in towns almost everywhere, although it has declined recently.

Length 15cm

| 1 | 2 | 3 | 4 | 5 |

J F M A M J J A S O N D

How to find it

■ **Timing** House Sparrows can be seen all year in the same areas. They form flocks in winter.

■ **Habitat** This species is closely associated with human habitation, and is common in most villages and towns, and around isolated farm buildings too, but has virtually disappeared from some city centres.

■ **Search tips** The male's loud advertising chirp in spring reveals the location of a nest-site, often within a crack or crevice in a house roof or gutter. Most gardens attract this species, especially if a seed mix is offered either in feeders or on bird tables.

WATCHING TIPS

House Sparrows are fun to watch as they feed, bathe and bicker together. In mid-summer adults can be watched feeding their newly fledged chicks. Avoiding pesticides in the garden will help ensure there are sufficient small insects for them to breed successfully.

Chaffinch

Fringilla coelebs

An extremely common bird of woodland and suburbia, the Chaffinch breeds throughout the UK and Ireland, except some Scottish islands.

Length 14cm

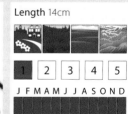

| 1 | 2 | 3 | 4 | 5 |

J F M A M J J A S O N D

How to find it

■ **Timing** Chaffinches are easy to find in woods and gardens all year. In winter many females migrate south of Britain, but numbers of males remain high.

■ **Habitat** This bird breeds in woodlands, scrub, hedgerows and well-vegetated gardens, and remains in similar areas through the winter. Numbers in upland woods and at rural feeding stations can be extremely high.

■ **Search tips** Listen for the bright 'pink' call and, in spring, the male's descending, chirruping song. Scan the forest floor for feeding Chaffinches, and watch out for them at your garden bird table, or picking up spillages on the ground under hanging feeders.

WATCHING TIPS

This is an attractive bird, with spring males very colourful. If you are lucky it may nest in your garden, providing great views in due course of the parents feeding the fledged babies. Substantial winter flocks in both town and country often attract other finch species.

Brambling

Fringilla montifringilla

This finch is the northern counterpart to the Chaffinch, breeding in huge numbers in the taiga forests of Eurasia, and visiting Britain in the winter months. Numbers vary between years, but are highest in the north and east.

Length 14cm

| 1 | 2 | 3 | 4 | 5 |

J F M A M J J A S O N D

How to find

■ **Timing** The first Bramblings may turn up in the far north in September, but the main arrival is through October and November. Most birds have departed by mid-April, although a handful hang on into May. It is an occasional breeder in north Scotland.

■ **Habitat** Bramblings wintering in Britain often flock with Chaffinches, and are generally found in similar habitat – woodland, hedgerows, weedy meadows, stubble fields and gardens. They are particularly keen on beechmast and when there is a good crop, beech woodland can attract large numbers of them.

■ **Search tips** Looking through Chaffinch flocks in winter is a good way to find Bramblings. Large flocks form where there are good feeding conditions, for example on stubble fields, in beech woods and in larger gardens if plenty of food is put out on the ground. In more northern parts Bramblings could even outnumber the Chaffinches. When scanning the flock, look out for the more variegated and orange-toned plumage of the Brambling, and if the flock takes flight, look out for white rumps.

WATCHING TIPS

Bramblings and Chaffinches are so similar in many respects that it is an interesting education to study them side by side and observe the ways in which they differ. The plumage patterns are different in enough small ways that identification is rarely a problem, but behaviourally there is much overlap. Bramblings in general seem to be rather more gregarious but less aggressive than Chaffinches, which ties in with their more migratory behaviour. A Brambling seen in mid-summer is a sign of local breeding and your local bird recorder should be informed.

Carduelis chloris

Greenfinch

A stocky, colourful finch, this familiar species is a very common and widespread resident throughout the British Isles.

Length 15cm

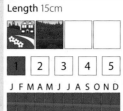

| 1 | 2 | 3 | 4 | 5 |

J F M A M J J A S O N D

How to find it

■ **Timing** Greenfinches are found in the same general areas all year except parts of the far north where they occur only in the breeding season.

■ **Habitat** These birds use wooded or partly wooded habitats with tall trees and an abundance of seed-bearing plants. They are common visitors to gardens in town and country.

■ **Search tips** In spring the Greenfinch's variable song, with twittered and nasal phrases, is delivered conspicuously from a high perch or an elegant, bat-like circling songflight. In the garden, look out for it on hanging nut or seed feeders.

WATCHING TIPS

Greenfinches are attractive garden visitors, which will use hanging feeders. They are, sadly, highly susceptible to the infectious and deadly disease trichomonosis – if you spot a sick-looking bird, take down and disinfect your feeders and encourage neighbours to do the same.

Carduelis carduelis

Goldfinch

This pretty bird is common and widespread everywhere except north-west Scotland, with the highest numbers in southern England.

Length 12–13cm

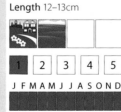

| 1 | 2 | 3 | 4 | 5 |

J F M A M J J A S O N D

How to find it

■ **Timing** Goldfinches are seen in open countryside all year, and are most likely to visit gardens in winter.

■ **Habitat** Unlike most garden birds, this species is more associated with open country than woodland. Look for it in scrubland, hedgerows, old orchards, weedy meadows, and drying-out marshland as well as gardens.

■ **Search tips** Goldfinches will visit most gardens if suitable food is offered. They are highly social – the tinkling calls announce the arrival of a flock or 'charm'. In late summer look for family groups feeding on large stands of seeding thistles and other tall weedy plants.

WATCHING TIPS

These delightful birds can often be tempted to a garden for the first time if you offer nyjer seeds in a specialised feeder – they also enjoy sunflower hearts. Dozens may visit at once. If you are lucky, adults may bring their fledglings to your garden from mid-summer.

Siskin

Carduelis spinus

This very attractive small finch, at home in woodlands and a regular garden visitor, is a common breeding bird across most of Britain and Ireland, except central and south-eastern England where it occurs as a common winter visitor.

Length 12cm

1	2	3	4	5

J F M A M J J A S O N D

How to find

■ **Timing** In areas where they breed, Siskins can be heard singing from early spring. They will come to gardens all year round. In the winter-only range, look out for them from late September through to April – many winterers come from the near continent and numbers increase in cold spells

■ **Habitat** These finches breed and overwinter mainly in mixed and coniferous woodland, with very high numbers in the native pine woods of the Scottish Highlands. In winter, flocks often gather in stands of alder trees in damp woodland. They are attracted to hanging feeders in gardens.

■ **Search tips** Male Siskins sing their pleasant twittering song from high in the treetops, where they can often be spotted outlined against the sky. In winter, flocks are very active, moving from tree to tree with a constant chorus of wheezing contact calls. When feeding, they are agile and hang upside down to access alder cones – in silhouette, the deeply forked tail distinguishes them from tit species. They may flock with redpolls, or with tits. Traditionally they were attracted only to red mesh bags of peanuts in the garden, but now you may see them using any hanging feeder.

WATCHING TIPS

Though small, Siskins are feisty birds and if they come to your garden you may find they dominate and monopolise the feeders, threatening even Great Tits and other larger species with an open-winged display. The males are brightly coloured and unmistakable, but females can be confused with other species at first glance. Their small size, streaky plumage and yellow patches help to identify them. Siskins may nest in your garden if you have tall conifers, while large feeding flocks make a noisy spectacle in winter.

This is a mostly brown but attractive medium-sized finch, common and widespread throughout Britain and Ireland except on the uplands of the Scottish Highlands. Its numbers are highest along the east coasts of England and Scotland.

Length 13.5cm

1	2	3	4	5

J F M A M J J A S O N D

How to find

■ **Timing** Linnets can be found in the same general areas year-round. In winter they are probably easier to see as they form large, active flocks.

■ **Habitat** You could find Linnets anywhere where there is open ground and scrubland or other bushy vegetation. Look on heathland and low-lying boggy moors, along farmland hedgerows, open rough grassland with tall weedy vegetation, on dry open scrub, in coastal fields with gorse bushes, and (especially in winter) on saltmarshes. In winter they may flock with other finches or (on the coast) with Snow Buntings, and occasionally visit gardens.

■ **Search tips** Linnets are highly vocal. In spring and summer, males sing a melodious twittering song from low but exposed perches. Flocks in winter call constantly, especially in flight. If you are walking in open countryside and encounter a noisy, tight-knit flock of sparrow-sized birds, there is a good chance they are Linnets – look out for the white flash in the primary feathers as they fly, with a distinctive undulating motion. On the ground they can be quite inconspicuous.

WATCHING TIPS

Listening to a male Linnet in song, it is easy to understand why this species was historically so popular as a singing cage bird. In breeding plumage the male is quite colourful, with a red crown and variable pink flush to the breast – although some males are nearly as drab as females. In all plumages, look for the white wing panel to identify a Linnet. This species is a rare visitor to gardens – most red-capped finches at the bird table turn out to be redpolls of one species or another – but if you live near suitable Linnet habitat you may be lucky.

Twite

Carduelis flavirostris

A more northerly relative of the Linnet, the Twite is a rather plain brown finch that breeds on coasts and uplands in Scotland and northern England and the west Irish coastline, and winters on some parts of the east coast.

Length 13.5cm

1	**2**	3	4	5

J F M A M J J A S O N D

How to find

Super sites

1. RSPB Balranald
2. RSPB Coll
3. Walney Island
4. Scammonden Dam
5. Saltfleetby NNR
6. RSPB Titchwell
7. Thornham Harbour
8. RSPB Old Hall Marshes

■ **Timing** Twites can be found in their coastal breeding areas all year, but only in spring and summer in the inland upland areas. They arrive at coastal wintering sites from October and stay until March – on migration they may stop off at inland sites.

■ **Habitat** Look for Twites in summer in areas of rough grassland with gorse or other bushes by the coast, and inland on grassy and heathy moorland. In the southern wintering areas they are usually found on saltmarsh, sometimes on vegetated shingle beaches or in coastal scrub. Passage migrants sometimes stop off on the shores of inland reservoirs.

■ **Search tips** When in suitable summer habitat, look out for Twites perched on the tops of bushes or on barbed wire fences. The male's song may draw your attention – it is quite harsh and emphatic for a finch, with twittered phrases. Twites also have a repertoire of rather Linnet-like calls. In winter, Twites form flocks and are usually found close to the sea. Linnets in winter plumage look rather similar – look out for the male Twite's pink rump in flight, and if you get a good view look for the yellow bill to confirm identification.

WATCHING TIPS

The Twite is our least colourful finch, but is still attractive in its rather muted way, with a rich warm cinnamon flush to its face and neatly streaked body plumage. Numbers on southern wintering sites in particular have fallen, and it is at these sites that it is most likely to be overlooked among the more numerous Linnets, so close examination of winter Linnets is worthwhile. The two species may flock together, allowing for side-by-side comparison – if they will keep still long enough.

This redpoll, also known as the Mealy Redpoll, is a rather uncommon but widespread winter visitor to Britain, most numerous in northern and eastern areas. It is usually seen with Lesser Redpolls, and numbers vary greatly from year to year.

Length 14cm

1	2	**3**	4	5

J F M A M J J A S O N D

How to find

■ **Timing** Common Redpolls begin to arrive in October, but most sightings are through the winter months. However, it is usually evident by the end of November whether it is going to be a good winter for the species.

■ **Habitat** These redpolls feed mainly on tree seeds through the winter months, and have a fondness for alder cones and birch seeds. They are therefore often found in damp wooded areas, but many sightings also come from gardens. They will nearly always be found flocking with Lesser Redpolls, and sometimes also with Siskins.

■ **Search tips** It is worth checking any redpoll flock between autumn and spring for this species. Look out for a larger, longer-tailed, paler and greyer bird. Some Common Redpolls are, however, much more obvious than others. Photographs can help you to study the key differences in structure and plumage. If your garden attracts Lesser Redpolls in winter then you stand a fair chance of seeing a Common Redpoll or two in a good year – look out for them on hanging feeders filled with sunflower hearts or nyjer seeds.

WATCHING TIPS

The redpolls that visit Britain represent a confusing variety of species and subspecies. Two different subspecies of Common Redpoll may occur, although the differences between them are usually too subtle to be noticed in the field. Then there is the much rarer Arctic Redpoll, again represented by two different subspecies. Years that are good for Common Redpolls are also usually good for Arctic Redpolls, so be aware of this possibility when checking a redpoll flock, and look out for any birds with pure white rumps and undertails.

Lesser Redpoll

Carduelis cabaret

This is a compact little woodland finch, which breeds mainly in Scotland, Wales and north-east England, but occurs in larger numbers in southern England through the winter. Some of our breeding birds overwinter in the near continent.

Length 11.5cm

1	2	3	4	5

J F M A M J J A S O N D

How to find

■ **Timing** In the breeding areas, Lesser Redpolls can be seen year-round, but will be most noticeable in early spring when the males begin to sing, and then again in winter when foraging flocks are moving through the woodlands.

■ **Habitat** The best habitat is damp deciduous, mixed or coniferous woodland, upland or lowland. Winter flocks are attracted to areas with a high proportion of birches and alders, and are frequently mixed with Siskins. The Lesser Redpoll is also a regular visitor to gardens with hanging seed feeders.

■ **Search tips** In winter, listen for the flight call, which is a two- or three-note rather hard rattling twitter, and check overhead for flocks on the move or feeding in nearby treetops. Often a few Lesser Redpolls will be found in a larger flock of Siskins, so check through all Siskin flocks carefully. Any red-crowned finch seen in woodland is likely to be this species rather than a Linnet. Singing males may be heard in breeding areas in spring – the song is a rather soft jumble of fast and sometimes harsh notes, including a rapid Wren-like rattling trill.

WATCHING TIPS

If you live in the country or suburbs and put out hanging feeders of sunflower hearts and nyjer seed, you stand a good chance of attracting Lesser Redpolls from late autumn through to early spring. This will give you an excellent opportunity to compare the various plumages of male, female and first-winter birds, as well as the chance to watch these engaging and lively little finches. They are Siskin-like in terms of their agility but are more peaceable and on large feeders many will feed together at the same time, alongside other species such as Goldfinches.

Loxia curvirostra

Common Crossbill

This chunky finch with its specialised bill for dismantling pine cones is a nomad. It does have some regular breeding areas but can occur and breed in pine woods anywhere following an 'irruption' (a mass arrival of birds from the continent).

Length 16cm

| 1 | 2 | 3 | 4 | 5 |

J F M A M J J A S O N D

How to find

■ **Timing** Common Crossbills can be seen all year in core areas. Irruptions generally begin in late summer or early autumn, and in the following months large numbers may be found in small or large pine woodland anywhere, or overflying open countryside as they search for places to feed.

■ **Habitat** These finches feed almost exclusively on pine seeds, all year round, and particularly favour the cones of spruce species. Large stands of conifers hold breeding populations, but wandering flocks will visit isolated copses of only a few trees, often within a heathland. Their diet means they require plenty of water – good habitat will have permanent pools or rutted ground where puddles can form.

■ **Search tips** The flight call of this species is a brief but very distinctive 'glip' and will reveal a party of thickset, fork-tailed finches going overhead. Look for feeding birds in any stand of pines (especially spruces) with plenty of cones – the birds will feed in flocks high in the trees, attacking the cones vigorously and keeping their balance with much fluttering. Also try watching woodland pools for birds coming to drink.

Super sites

1. RSPB Corrimony
2. RSPB Abernethy Forest
3. Coronation Plantation
4. Forest of Dean
5. Breckland
6. RSPB The Lodge

WATCHING TIPS

This species is a very interesting bird, with the capacity to breed at any time of year in order to take advantage of good food supplies. In irruption years, the birds that arrived at the end of summer may begin to nest the following January, although they and their offspring may then move back to the continent. With good views, the crossed mandibles can be seen, and feeding birds can be watched using this unique feeding equipment to expertly extract seeds from spruce cones. Males vary in colour from orange to vivid red, while females may be greenish or more yellow.

Scottish Crossbill

Loxia scotica

The Scottish Crossbill has the distinction of being Britain's only endemic bird species, occurring nowhere else in the world. It is confined to the Scottish Highlands, and its similarity to other crossbills presents birdwatchers with a major identification challenge.

Length 16.5cm

1	2	3	4	5

J F M A M J J A S O N D

How to find

■ **Timing** Scottish Crossbills are generally quite sedentary, but will wander short distances when necessary to find food. They may be easier to see at ponds after a dry spell, when there will be fewer temporary pools and puddles where they can drink.

■ **Habitat** Like the Common Crossbill, this is a bird of pine forest, especially the native Scots pine woodland of Speyside and Deeside, but also in plantations, especially those of larch and lodgepole pine. It feeds on the seeds of Scots pine and larch in preference to the various spruce species, and drinks frequently so needs easy access to fresh water.

■ **Search tips** Look for this crossbill feeding on Scots pine cones up in the trees, coming down to the forest floor to drink from ponds or puddles, and flying overhead, giving a clipped metallic call that differs slightly from the Common Crossbill's usual flight call. It is usually found in small parties, and may be seen alongside both Common and Parrot Crossbills – in the Scottish Highlands it is not possible to make assumptions about any crossbill's identity just by its location.

Super sites

1. RSPB Abernethy Forest
2. Grantown Woods

WATCHING TIPS

Finding crossbills in the Highlands is not all that difficult, but unfortunately their identification is very difficult. The Scottish Crossbill sits in between Common and Parrot Crossbills in terms of general size and bill size, so it is very difficult to be sure that you have not found a big, large-billed Common or a small, smaller-billed Parrot. For those seriously interested in crossbill identification, recording the birds' calls and analysing graphic representations of their waveforms (sonograms) is the surest way to make a confident identification.

Loxia pytyopsittacus

Parrot Crossbill

This large crossbill is a very rare breeding bird, found only in Strathspey pine woods, and an unpredictable rare visitor.

Length 17.5cm

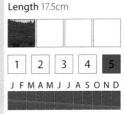

| 1 | 2 | 3 | 4 | 5 |

J F M A M J J A S O N D

How to find it

- **Timing** Parrot Crossbills in Abernethy Forest may be seen all year. If the pine crop in Scandinavia fails, there may be an irruption with sightings elsewhere, at any time.
- **Habitat** Like other crossbills, this bird feeds mainly on pine seeds so is found in coniferous forest – primarily in Abernethy Forest in Strathspey.
- **Search tips** The same search advice as for the other crossbills applies here – the species is usually seen feeding on cones or drinking at forest pools. You may notice its flight call, which is deeper than that of the other species.
- **Super sites** 1. RSPB Abernethy Forest.

WATCHING TIPS

Adult male Parrot Crossbills can be quite distinctive, as crossbills go, with their heavy bills and thick 'bull-necks'. However, identifications should always be made with great caution. Parrot Crossbills away from the Scottish Crossbill breeding area are easier to identify.

Carpodacus erythrinus

Common Rosefinch

This finch is a scarce passage migrant, mainly in the east, and sometimes one or two stay for the summer.

Length 14.5–15cm

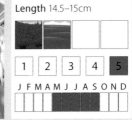

| 1 | 2 | 3 | 4 | 5 |

J F M A M J J A S O N D

How to find it

- **Timing** Most sightings are in May and September, but late migrants or oversummering birds may be seen in between these months.
- **Habitat** Common Rosefinches are usually found at or near the coast, in areas with bushes and some trees. If they make landfall in more open areas, they will use whatever cover they can find.
- **Search tips** Males will sometimes hold territory and give their pleasant song in spring (occasionally through into summer), using exposed perches. These birds may be colourful, with scarlet heads and breasts, or much duller. Juveniles in late summer are shyer, hiding in thick cover.

WATCHING TIPS

The Common Rosefinch has long been expected to begin to colonise the UK, but so far breeding has been occasional and sporadic. On the east coast especially, look out for this bird throughout summer as any evidence of breeding is significant and should be passed on to your local recorder.

Bullfinch

Pyrrhula pyrrhula

A beautifully coloured, rather shy finch, this bird is common and widespread in Britain and Ireland except for the far north-west of Scotland and some of the northernmost islands. It is a fairly common visitor to more rural gardens.

Length 14.5–16.5cm

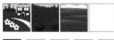

1	2	3	4	5

J F M A M J J A S O N D

How to find

■ **Timing** You can see Bullfinches in the same areas all year, and in gardens more often between late autumn and early spring. The distinctive north-eastern European subspecies is a scarce and sporadic winter visitor.

■ **Habitat** Bullfinches like light deciduous woodland and scrubby areas on the edges of woods, with plenty of bushes or small trees that produce blossom in spring and fruit in autumn. They are often found in orchards, as well as quiet parkland and large, well-planted gardens.

■ **Search tips** This is a quiet and discreet bird, quick to disappear when it sees you coming. Listen for the call, which is a soft and brief, but very distinctive, plaintive whistled 'piu', and when you hear it scan through the tops of trees for the birds, which are often in pairs. A Bullfinch in retreat shows a flash of white rump, and when the sun catches the male's pink breast it can stand out from a distance. Feeding stations in wooded or woodland edge areas often attract Bullfinches, especially if they are screened off so there is minimal disturbance.

WATCHING TIPS

Because these birds are shy and flighty, the easiest way to watch them is from the kitchen window – try offering mixed seed scattered on a bird table that is close to or suspended from an overhanging tree to provide shelter. Unusually, the pair bond persists throughout the winter months, so you are unlikely to attract them in large numbers, but if they breed locally you should see juvenile birds from mid-summer. Birds of the northern subspecies, larger and brighter than our breeding birds, sometimes occur in significant numbers in winter – listen for the distinctive 'toy trumpet' call.

Coccothraustes coccothraustes

Hawfinch

This beautiful, striking but very shy woodland finch is a rather uncommon breeding bird in Britain, found mainly in England and Wales but just reaching into south-eastern Scotland, with the highest numbers in southern England. It is absent from Ireland.

Length 18cm

| 1 | 2 | **3** | 4 | 5 |

J F M A M J J A S O N D

How to find

■ **Timing** Although Hawfinches are on their territories all year round, they are much easier to see in winter and early spring, when they form small flocks and are more noticeable in the leafless trees.

■ **Habitat** This finch is found in mixed and deciduous woodland that is not much disturbed by human activity. It has a particular affinity with hornbeam trees, and feeds on tree seeds like cherry and plum that are too hard for other finches to crack. It roosts (sometimes communally) in conifers.

■ **Search tips** Its very shy temperament makes this a difficult bird to find – if it sees you coming it will silently slip out of sight. When walking in suitable woods in autumn and winter you may find it feeding on beechmast and other fallen tree seeds on the ground – walk as quietly as you can and scan well ahead of your path. Listen also for the flight call, a dry abrupt 'tick'. Winter flocks may assemble in treetops before going to roost in the shelter of conifer trees. Although a regular visitor to gardens and feeding stations on the continent, it is not a garden bird in Britain.

Super sites

1. Scone Palace
2. Woodwell
3. Clumber Park
4. Wyre Forest
5. RSPB Nagshead
6. Lynford Arboretum
7. Pitts Wood
8. Bedgebury Pinetum

WATCHING TIPS

You'll need patience and luck to watch these spectacular, huge-billed finches. At even the best sites views may be limited to silhouettes in high treetops against a darkening winter sky. If you have the time, resources and permission of the landowner, setting up a temporary hide in a suitable area where there are plenty of the right kinds of trees could pay dividends. It may also be worth staking out a woodland pond (especially if there has been dry weather and there are few puddles around) as these finches are frequent drinkers.

EXPLORING THE COUNTRY

When you think of a birdwatching outing, there's a good chance you are envisaging a trip to a nature reserve, with its trails, information boards and well-placed hides. However, we should not fall into the trap of thinking that birds can only be seen on reserves – or it could become a self-fulfilling prophecy. Protecting birds in the wider countryside is as important a part of conservation as is establishing nature reserves, but if ordinary birdwatchers don't go there, it becomes easy to assume that vast tracts of countryside have no birds worth protecting.

Public and private

One of the key investments you should make as a new birdwatcher is a copy of your local Ordnance Survey map. Even if you live in an area replete with nature reserves, it is still well worth finding out where your local public footpaths lie and going out for some serious exploration. The map will show you how the terrain changes from flat to hilly, what general habitat types you'll pass through, and most importantly how to plan your route along public rights of way.

It is very important to keep off private land unless you have permission to be there. While some landowners will be relaxed about wandering walkers, others will be justifiably annoyed. In due course, you may find yourself approaching landowners to request their permission for access to their land, and so it will pay to get any relationship off on the right foot. Many farmers and other landowners have a strong interest in wildlife and will happily allow you access to private land provided your approach is respectful, you can assure them that you know how to conduct yourself and that your interest is genuine.

Making the most of it

On a nature reserve, the hard work is often done for you. A sightings board at the entrance lists the birds that have been seen, birdwatching hides give you the best possible views of the best areas, and there may even be volunteers around to point out the exact birds that you're looking for. In the countryside, it's down to you to do the finding.

The first lesson to learn is to walk slowly. It is easy to get into the 'quick march' mentality of an invigorating country walk, but if you walk too fast you will disturb more birds than you'll see. So slow it down, and pause often to look around. Good places to stop are where one type of habitat meets another. Pick a route that incorporates lots of borders – if you have the option of walking across the middle of a field or along the edge where it adjoins a woodland, go for the edge.

Look out for spots that give you a wide view. An elevated position looking down onto a woodland can be excellent. On the beach, try to choose a path from where you can see the sea but also the land on the other side. Consider light direction too – pick a route where the sun will be behind you for the most interesting parts of the walk, so the birds you see will be well lit rather than silhouetted.

The reward of this kind of exploration is the discovery of a really productive 'local patch' of your own. Keep notes of your sightings – they may prove invaluable if your patch should come under threat of development. The RSPB and other conservation bodies are always looking for valuable wildlife habitat to protect for the future.

PERSONAL SAFETY

You are most unlikely to come to any kind of grief on a nature reserve with all the facilities. Once you venture into the wider countryside, though, things may be different. When you walk a new route for the first time, be particularly alert for potential hazards, and go prepared. Take a mobile phone with you, and a bottle of water – even if your planned route is short, you may be forced to double back if the path is obstructed, and end up walking much further than you intended.

The vast majority of people you'll encounter while out walking will be pleasant and like-minded souls. However, it may be sensible to leave large, expensive optics at home the first few times you walk an unfamiliar path, especially on the outskirts of a town, until you feel confident that you are unlikely to meet less agreeable people on a regular basis.

Lapland Bunting

Calcarius lapponicus

This handsome bunting breeds in north Scandinavia and further east, and migrates south in winter. It is a scarce winter visitor mainly to the east coast of Britain, and frequently associates with Snow Buntings when it is here.

Length 15–16cm

1	2	3	4	5

J F M A M J J A S O N D

How to find

■ **Timing** Although the first Lapland Buntings arrive in September and the last to leave linger into May, the period between November and March is the best time to look for this species.

■ **Habitat** Most Lapland Buntings in the UK stay on or very near the coasts, where they look for seeds and other plant matter on vegetated shingle, tidelines, rough grassland, dunes, and in coastal arable fields. There are usually more records in northern areas.

■ **Search tips** This bird is usually encountered when walking a coastal path in a low-lying area. Large Snow Bunting flocks often attract a few Lapland Buntings, so if you see any Snow Buntings carefully check each bird – the lack of any large area of white in the plumage will help you pick out a Lapland Bunting. Also listen for the call, a hard rattling short trill ending with a sweeter 'peu' note that stands out among the musical jingling twitter of the Snow Buntings. Lapland Buntings are also found on their own – when walking, pause to scan fields and shorelines, as the birds are quite inconspicuous when feeding on the ground.

Super sites

1. RSPB Dee Estuary – Point of Ayr
2. Teesmouth
3. Filey Brigg
4. Donna Nook
5. RSPB Frampton Marsh

WATCHING TIPS

These attractive and charismatic birds, like Snow Buntings, are usually quite confiding and approachable and may not take flight even if you pass by quite close to them, although flocks can be restless, taking off and moving to a new area for no obvious reason. Lone birds could be mistaken for Reed Buntings, but their different habits are as apparent as the differences in plumage and structure. Late-staying birds (more likely in the north) may develop full breeding plumage before departing, which in the male is very striking and attractive.

Snow Bunting

This charming and beautiful bird is a rare breeding species in Britain, found only on the highest mountains in Scotland. It also occurs along lowland coasts as a winter visitor, and is most regular and numerous in the north.

Length 16–17cm

| 1 | 2 | 3 | 4 | 5 |

J F M A M J J A S O N D

How to find

■ **Timing** Between May and August, Snow Buntings are on their breeding grounds. From September to April they move to low ground, mainly on the coast, and numbers in mid-winter may be augmented by arrivals from the continent.
■ **Habitat** The breeding habitat is exposed and barely vegetated mountain tops, well above the tree line and often alongside permanent snow patches, where it nests in sheltered spots among boulders. Winter habitat is along flat coastlines and includes vegetated beaches (shingle or sandy), saltmarsh, coastal grazing marshes and arable fields.
■ **Search tips** This is the only small bird you are likely to encounter on the mountain tops and is usually obvious and approachable – look for it in dips and crannies that are sheltered from the wind, and anywhere where there are small patches of vegetation. In winter, scan fields and beaches, bearing in mind that on shingle in particular it is surprisingly well camouflaged. Flocks often make short flights between feeding spots and are obvious in flight with their sweet-toned calls and flashing white wings. Further south, you are more likely to find them in small groups or even as single birds.

Super sites

1. RSPB Balranald
2. Cairngorms
3. Carn Ban Mor
4. RSPB Lough Foyle
5. Filey Brigg
6. Saltfleetby NNR
7. RSPB Frampton Marsh
8. RSPB Titchwell
9. Cley Marshes WT
10. Sandwich Bay

WATCHING TIPS

Whether you meet them in winter or summer, Snow Buntings are easy to watch as they have little fear of people. Many walkers in the mountains have stopped for lunch and had Snow Buntings at their feet, picking up sandwich crumbs. As long as you keep reasonably quiet and still, you should enjoy close views, although winter birds are restless and may disappear suddenly to visit another part of their extensive feeding range. Always check Snow Bunting flocks for both Lapland Buntings and Shore Larks.

Yellowhammer

Emberiza citrinella

The Yellowhammer is a common and widespread bird (although it has suffered serious declines). It is found throughout most of the British Isles but is most common in southern and eastern areas, and on the east side of Ireland.

Length 16–16.5cm

| 1 | 2 | 3 | 4 | 5 |

J F M A M J J A S O N D

How to find

■ **Timing** You can see Yellowhammers all year round in most areas, although they move away from northerly uplands in winter. They are also most likely to be found in gardens in winter.

■ **Habitat** Yellowhammers can be found in a variety of dry, mainly low-lying open habitats with scrub. Good kinds of habitat for the species include quiet mixed farmland with plenty of mature hedgerows, coastal fields with stands of gorse, and heathland. They mainly visit more rural gardens adjoining open countryside, but in cold spells may turn up in suburbia.

■ **Search tips** A very distinctive rattling song with a long single 'tweeze' note at the end (written as 'a-little-bit-of-bread-and-no-cheese') reveals the presence of a male Yellowhammer. When you hear it, scan the tops of bushes and small trees. The male's vivid yellow face and breast, contrasting with rufous-brown body, is usually quite eye-catching. Yellowhammers flying along hedgerow tops look long-tailed, and the chestnut rump is conspicuous. In winter, look for Yellowhammers feeding inconspicuously on the ground in stubble fields, among grain spills on field edges, and at feeding stations on woodland edges.

WATCHING TIPS

Yellowhammers are quite easy to watch, although they are not especially approachable so you will need to keep your distance to avoid disturbing them. Males pick prominent song perches and may sit still singing on the same perch for very long spells unless disturbed. Females and juveniles are drab birds by comparison, but still a welcome sight, whether encountered in the wider countryside in mixed winter flocks with finches and other buntings, or in your garden. To attract this species, offer seed on the ground or on bird tables.

Cirl Bunting

Once quite a common bird in southern England, the Cirl Bunting's range has contracted over the last century to a small area on the south coast of Devon. Careful habitat management is now helping it to spread from this stronghold.

Length 15.5cm

| 1 | 2 | 3 | **4** | 5 |

J F M A M J J A S O N D

How to find

■ **Timing** Cirl Buntings can be seen at the same sites all year round. They are probably easier to see in spring when males are singing, or in winter when they form flocks, and the leafless hedges and bushes offer them fewer hiding places.
■ **Habitat** The UK population of Cirl Buntings frequent areas of farmland, at or near the coast, with arable fields and hedgerows. There are several sites in Devon where winter stubble fields are left specifically for the birds to use, and flocks form at these places (with other seed-eating birds in tow).
■ **Search tips** In spring, listen for the male's song, which is a rattled trill like that of a Yellowhammer only without the final long note (more similar, in fact, to the song of the Lesser Whitethroat). When you hear it, scan the tops of hedgerows, fence posts and along fence wires and even telegraph poles, as males sing from exposed and usually elevated positions. Singing is more likely on still, fine days. In winter, find spots in the stubble fields where finches and buntings have gathered and check through them for a male Cirl Bunting (females are extremely difficult to distinguish from female Yellowhammers).

Super sites

1. RSPB Labrador Bay
2. Berry Head
3. Prawle Point

WATCHING TIPS

If you live in Cirl Bunting country, you may be able to attract them to your garden by offering a seed mix at ground level. This will give you a good opportunity to study the plumage in detail – unmistakable in males, but potentially confusing in females. When faced with a possible female Cirl Bunting, the most important thing to check is the rump colour – a cool grey brown in Cirl Bunting, but a brighter rufous in Yellowhammer. Cirl Buntings are spreading from their core area of Devon and are being reintroduced to Cornwall, so look out for them in neighbouring areas.

Reed Bunting

Emberiza schoeniclus

This quite conspicuous small bird of wetland and scrubby areas is a widespread resident species, common throughout the British Isles. It is also a surprisingly frequent visitor to gardens in winter, especially in spells of severe weather.

Length 15.5–16.5cm

1	2	3	4	5

J F M A M J J A S O N D

How to find

■ **Timing** Reed Buntings are easy to see in spring when the males are singing. In winter, some move away from wetlands to drier habitats, and this is also the likeliest time for them to visit gardens.

■ **Habitat** Breeding habitat is sheltered wetland areas, with reedbeds, bulrushes, sedges, bushes and other tall vegetation offering safe nest-sites. In winter some birds will remain in wetland areas, but others will visit stubble fields, open scrubland, hedgerows and rural and suburban gardens. It is most likely to abandon the wetlands when there is a freeze.

■ **Search tips** The male's simple, monotonous three-note song is given through spring and summer. Like other buntings, it sings from quite high and exposed perches, which stand out particularly well in a rather homogenous wetland habitat – if there is a lone small bush in the reedbed, this is the first place to check for the singing bird. When not singing, though, it can keep a very low profile among the reeds. In winter, carefully check through all flocks of finches, buntings and sparrows that you see – females and winter-plumaged males are easily missed among House Sparrows.

WATCHING TIPS

Although it is quite easy to watch singing Reed Buntings, their lives in the breeding season beyond this can be quite mysterious, given the nature of their habitat. If you notice an adult bird carrying food, there will be a nest nearby and there could be fledglings to see in a few days' time. You will have a great chance to watch Reed Buntings if they visit your garden – the cold snaps in the winters of 2009–2010 and 2010–2011 brought them in large numbers to gardens throughout Britain, even quite built-up areas. They are most likely to feed on seed on the ground or on bird tables, but sometimes learn to use feeders.

Emberiza calandra

Corn Bunting

A large and heavy-set bunting, this species has suffered serious declines and local extinctions (including from Ireland) over the last few decades. In England, eastern Scotland and eastern Wales it remains quite widespread and in some areas common.

Length 18cm

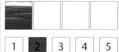

1	2	3	4	5

J F M A M J J A S O N D

How to find

■ **Timing** Corn Buntings can be seen all year round. The males are easiest to find when singing in spring and into summer, while winter feeding flocks form from autumn.

■ **Habitat** This is a true farmland bird, found in lowland areas of arable and mixed farming where there are hedgerows or other scrubby areas for cover, and thick ground vegetation for nesting. It is often found in more open areas with larger fields than other open-country buntings, meaning that it tends to be encountered in single-species flocks.

■ **Search tips** The song of this bird is sometimes described as resembling the jangling of a bunch of keys. Like the songs of other buntings, it is distinctive, and usually delivered from an obvious elevated spot. When you hear it, scan the tops of the hedges, telegraph poles, overhead wires, tall weedy plants and any small trees around. Winter flocks can be quite sizeable, and if disturbed from where they are feeding on the ground the birds will often fly up to settle on overhead wires.

WATCHING TIPS

Singing male Corn Buntings will provide you with a constant background chorus of song when you are walking in suitable farmland in spring, and they can sometimes be approached quite closely. When arriving at or leaving a song post, they hover momentarily with legs dangling. Meanwhile, the female or females (the species is sometimes polygamous) will be out of sight at the base of a hedge or in weedy vegetation, tending the nest with no assistance from the male. Its declining numbers and shrinking range means that monitoring its status is very important.

Glossary

Courtship – a set of behaviours performed by a pair of birds before mating

Display – any behaviour pattern that is performed to convey a message to another bird or animal

Distraction display – a set of behaviours performed by a bird with the intention of leading a predator away from the nest, e.g. feigning a broken wing

Down-slurred – a term used to describe a bird call that 'slides' from a higher to a lower pitch without being two separate notes

Eye-ring – a prominent circle of bare skin around a bird's eye, usually a colour that contrasts with the surrounding feathers

Hirundine – a bird of the swallow and martin family

Irruption – an unpredictable mass movement of birds in response to a resource shortage

Juvenile – a bird in its first set of feathers, which may be retained for just a few weeks or for most of its first year of life

Leucism – Abnormal pale or white feathers in a bird's plumage – may be partial or total

Migrant – any bird that migrates

Migration – a regular seasonal journey from one area to another

Optics – equipment for birdwatching; binoculars and telescopes

Passage migrant – a bird that visits us only when on its migratory journey, not staying to breed or to overwinter

Passerine – a bird belonging to the taxonomic group Passeriformes, also known as songbirds and perching birds

Primaries – the long wing feathers on the outer part of a bird's wing

Seawatching – looking for and watching birds at sea

Secondaries – the long wing feathers on the inner part of a bird's wing

Skydancing – a ritualised courtship flight, often very dramatic, performed by one or a pair of birds of prey

Songflight – a stylised flight pattern performed by a singing bird, e.g. a Skylark

Supercilium – A contrastingly coloured stripe above the eye, usually pale

Territory – an area that a bird or birds defends against other members of its species, usually used for breeding but also sometimes just for feeding

Tertials – the innermost three long wing feathers

Wader – a sandpiper, plover or allied species

Wing-bar – a contrastingly coloured band across a bird's wing

Acknowledgements

Thanks to Nigel Redman at Christopher Helm for backing this project, and Lisa Thomas, Julie Bailey and Sarah Cole for seeing things through to publication with good humour and much patience. It was a pleasure to work once again with the talented Julie Dando, who produced the book's design and compiled the index, and to work for the first (but hopefully not the last) time with Sara Hulse, whose skilled copyediting streamlined the text and weeded out one or two embarrassing author errors. Thanks also to Toni Tochel for proofreading, and the many photographers who supplied images for the book.

As ever, I'd like to thank Rob for his patience and indulgence, for taking some of the lovely photos that have appeared in this book, and making sure I was never in want of a cup of tea on the long evenings of writing. I must also thank the many friends and a few relations who have joined me 'in the field' for birdwatching adventures up and down the country, especially those who were not birdwatchers at all… but out of kindness indulged my interest and ended up developing a fledgling interest of their own. I am dedicating this book to my best buddy Michèle, who has always exemplified this generous and open-minded attitude. When she texted me to say I should hurry round because the House Sparrows were feeding their chicks in her garden, I knew the birds had won her over.

Recommended reading

The following books and websites are recommended for everyone with an interest in birds in general, and bird-finding in particular.

Books

Where to Watch Birds in Britain by Simon Harrap and Nigel Redman (Christopher Helm). A detailed gazetteer of all of the best birdwatching sites in Britain.

RSPB Where to Discover Nature by Marianne Taylor (Christopher Helm). A visitor's guide to the UK's RSPB reserves.

The Ultimate Site Guide to Scarcer British Birds by Lee Evans (Birdguides). Comprehensive site information for the rarer British birds.

RSPB Handbook of British Birds by Peter Holden and Tim Cleeves (Christopher Helm). An in-depth field guide to all regularly breeding and visiting British birds.

Collins Bird Guide by Lars Svensson, Killian Mullarney and Dan Zetterström (Collins). A field guide to birds of Britain and Europe, including all regular, rare and accidental species on the British list.

Websites

www.rspb.org.uk Your one-stop resource for the RSPB. Includes a species guide, reserves listings, a community zone with message boards and reserve wardens' blogs, conservation news, and details of the RSPB's work, past and present.

www.bto.org The website for the British Trust for Ornithology, home of BirdTrack and other bird surveying projects.

www.wwt.org Online home of the Wildfowl and Wetlands Trust, a charity dedicated to conserving wildfowl and their habitat.

www.birdguides.com A comprehensive resource for birdwatchers, with news of sightings (in detail for subscribers), trip reports, a products shop, photo galleries and more.

www.rarebirdalert.net Provides up-to-date details of rare bird sightings for subscribers (non-subscribers can view basic information).

www.birdforum.net This is a busy online community for birdwatchers to exchange news, seek advice, show off photographs and chat. Also includes a global wiki on birds and birdwatching sites.

www.wildaboutbritain.co.uk Information and a collection of discussion forums on birds and other British wildlife.

Index

Bloomsbury Publishing would like to thank the following for providing photographs. While every effort has been made to trace and acknowledge all copyright holders, we would like to apologise for any errors or omissions, and invite readers to inform us so that corrections can be made in any future editions of the book. T = top, B = bottom, L = left, R = right and SS = Shutterstock.

Front cover: top Mike Lane/FLPA; bottom left Steve Ellis/SS; bottom middle ArvydasS/SS; bottom right Mircea Bezergheanu/SS. Spine: Markus Varesvuo. Back cover photos: David Tipling, except second top Accent/SS.

1 David Tipling/FLPA; 3 TTPhoto/SS; 5 Mark Sisson/FLPA; 6 David Tipling/FLPA; 8 Graham Bloomfield/SS; 9 RedTC/SS; 10T Tropper/SS; 10B Clinton Moffat/SS; 11T Markus Varesvuo; 11B Ran Schols/AGAMI; 12 Andrew Parkinson/FLPA; 13 Nobutoshi/SS; 14T Marianne Taylor; 14B Gerrit de Vries/SS; 15 Sebastian Knight/SS; 16 Jon Brackpool Photography/SS; 17 Ainars Aunins/SS; 18L Derek Middleton/FLPA; 18R Elliotte Rusty Harold/SS; 19T Malcolm Schuyl/FLPA; 19B Martin Fowler/SS; 20 Nick Biemans/SS; 21 Marianne Taylor; 22 David Dohnal/SS; 23 Mondegofoto/SS; 24 Martha Marks/SS; 25 Roger Tidman/FLPA; 26T Michael Quinton/Minden Pictures/FLPA; 26B Marianne Taylor; 27T Maslov Dmitry/SS; 27B Andrew Parkinson/FLPA; 28T Bill Coster/FLPA; 28B Robert Cardell; 29 Tom Vezo/Minden Pictures/FLPA; 30T Friedhelm Adam/Imagebroker/FLPA; 30B Martha Marks/SS; 31 David Hosking/FLPA; 32 Elliotte Rusty Harold/SS; 33 Paul Hobson/FLPA; 34 Markus Varesvuo; 35 Marco Barone/SS; 36 Dietmar Nill/Minden Pictures/FLPA; 37 Hugh Lansdown/FLPA; 38T Daniel Prudek/SS; 38B Robin Chittenden/FLPA; 39 David Tipling/FLPA; 40 Andrew Parkinson/FLPA; 41 Mike Lane/FLPA; 42 Michael Callan/FLPA; 43 Gary K Smith/FLPA; 44L KAppleyard/SS; 44R Ewan Chesser/SS; 45T Jane Rix/FLPA; 45B Wayne Hutchinson/FLPA; 46T Richard Brooks/FLPA; 46B David Dohnal/SS; 47 John Hawkins/FLPA; 48T Rob Kemp/SS; 48B Robin Chittenden/FLPA; 49 Mike Lane/FLPA; 50 Andrew Parkinson/FLPA; 50B Robin Chittenden/FLPA; 51T Hugh Lansdown/SS; 51B Iliuna Goean/SS; 52T Pavel Cheiko/SS; 52B Elliotte Rusty Harold/SS; 53T Razvan Zinica/SS; 53B Nigel Dowsett/SS; 54T Georgios Alexandris/SS; 54B Robin Chittenden/FLPA; 55 Martin H Smith/FLPA; 56T Leo Vogelzang/FN/Minden/FLPA; 56B Mike Lane/FLPA; 57T Mircea Bezergheanu/SS; 57B Razvan Zinica/SS; 58T Serg Zastavkin/SS; 58B David Thyberg/SS; 59T Daniele Occhiato/AGAMI; 59B Marc Guyt/AGAMI; 60T James Lowen/FLPA; 60B Robin Chittenden/FLPA; 61 Steve Young/FLPA; 62 Roger Tidman/FLPA; 63 Bill Coster/FLPA; 64 Steve Young/FLPA; 65T T. W. van Urk/SS; 65B Marianne Taylor; 66 Stephen Meese/SS; 67T Andrew Parkinson/FLPA; 67B David Tipling/FLPA; 68T Godrick/SS; 68B Georgios Alexandris/SS; 69T Arto Hakola/SS; 69B Gertjan Hooijer/SS; 70 Razvan Zinica/SS; 71T Arto Hakola/SS; 71B Gertjan Hooijer/SS; 72 Marianne Taylor; 73T Florian Andronache/SS; 73B Marianne Taylor; 74 Florian Andronache/SS; 75T Mircea Bezergheanu/SS; 75B BogdanBoev/SS; 76 Foto Bouten/SS; 77 Otto Plantema/Minden Pictures/FLPA; 78 RubberSideUp/SS; 79 Toomasili/SS; 80 P. Schwarz/SS; 81 Martin Woilke/FN/Minden Pictures/FLPA; 82 Roger Tidman/FLPA; 83 Steve Byland/SS; 84T Mircea Bezergheau/SS; 84B Papkin/SS; 86L Schaef71/SS; 86R Dickie Duckett/FLPA; 87T Tony Hamblin/FLPA; 87B Kippy Spilker/SS; 88 Neil Bowman/FLPA; 89 Harry Taavetti/FLPA; 90 Michael Callan/FLPA; 91T Mircea Bezergheanu/SS; 91B Menno Schaefer/SS; 92 Roger Tidman/FLPA; 93 Xpixel/ Shutterstock; 94T Paul Hobson/FLPA; 94B Mark Bridger/